lampshades&lighting

GLOUCESTER MASSACHUSETTS

ROCKPORT PUBLISHERS

lampshades&lighting

The Project and Decorative Sourcebook

First published in the United States of America by
Rockport Publishers, Inc.
33 Commercial Street
Gloucester, Massachusetts 01930-5089
Telephone: (978) 282-9590
Fax: (978) 283-2742
www.rockpub.com

ISBN 1-56496-926-6

10 9 8 7 6 5 4 3 2 1

Grateful acknowledgment is given to Lisa Skolnik for her work from *The Right Light* on pages 8–142; to Maryellen Driscoll for her work from *The Paper Shade Book* on pages 144–235, 276–289, and 292–293; to Mary Ann Hall for her work from *Paper House* on pages 236–245 and 274 and for her work from *The Crafter's Project Book* on pages 254–261 and 290–291; to Sandra Salamony for her work from *The Crafter's Project Book* on pages 254–261 and 290–291; to Lynne Farris for her work from *Baby Crafts* on pages 262–265; to Gail Hercher for her work from *Crafting with Handmade Paper* on pages 266–273; and to Randall Whitehead for his work from *The Art of Outdoor Lighting* on pages 294–297.

Cover Images: Kevin Thomas, (top left and bottom right)
                Elizabeth Whiting & Associates, (bottom left)
                Fritz von der Schulenberg/The Interior Archive (top right)

Printed in China

C ONTENTS

Introduction

Lighting is one of the most routine features of our homes—so routine, in fact, that we usually take it for granted. Most of us live without second thoughts with the standard-issue, built-in lighting that comes with our abodes. Of course, we supplement the overhead lighting in each room with lamps and fixtures of every ilk, yet we often don't choose these sources of light for the right reasons. They may look good, or solve an immediate problem by providing the illumination that is needed for a specific task, but are they really the right light?

Just as we carefully weigh the style of furniture or the palette of colors we use in our homes, we must carefully weigh the way we light them. Lighting does far more than merely illuminate a room; it also sets the stage for every act that takes place in it. Besides emphasizing areas of importance or activity, or

left
The stainless-steel hanging lights match the sink, stove hood, and refrigerator in this young and modern kitchen.

highlighting prized possessions, light can create the right ambience for a space, be it prosaic or dramatic. A beautiful room can fall flat if it is outfitted with inadequate lighting. The same is true of beautiful fixtures or lamps; they are worthless as light sources if unsuited to the tasks at hand.

The environment and ambience we establish in a space by lighting it a certain way also has a profound effect on many aspects of our psyche. Take efficiency and performance. We are more likely to be alert, energetic, and ready to work on a sunny day, which causes bright highlights and crisp shadows in a room. The opposite holds true on a drab day, when there is no contrast and that same room assumes a monotonous tone that lulls its occupants into lethargy. In essence, the difference between these two days is due to variations in the quality of natural light. Fortunately, artificial illumination can create contrasts in a room to simulate natural light, if necessary, and provide a stimulating environment.

Light does far more than provide the illumination we need for efficiency. It also affects our psychological well-being through the moods it creates; we feel calm and serene, industrious and alert, lively and enthusiastic, or romantic and relaxed. And it isn't just natural light that produces this effect. Making a significant impact with lighting is an intricate undertaking that calls for a careful balance of both natural and artificial light.

Yet lighting has long been underutilized as a home design tool for a number of reasons. For starters, it requires more planning than many other decorating tools. It is easier to settle for an overhead fixture and a few plug-in lamps than to implement more effective lighting treatments that require special wiring or structural changes. Also, lighting doesn't offer any obvious type of gratification; it is a subtle element that must be appreciated at the time of day or night when it is at work. Finally, lighting has not received the attention it deserves and requires simply because many homeowners are unaware of the impact it makes in a space and feel inadequate using it as a decorating tool. Instead of tackling the subject themselves—which can require some technical reading—or calling in a pro, homeowners ignore the challenge altogether.

While the generic options suggested above—simply adding ceiling fixtures and a smattering of lamps—have long been the default lighting solution in many rooms, these options are no longer a given. Savvy architects, designers, and do-it-yourselfers are finally considering lighting as a critical component of good home design. As well it should be; successfully balancing the functional and aesthetic aspects of lighting takes consideration, planning, and careful implementation on a room-by-room basis. Lighting should, and will, vary greatly between rooms, depending on its function in each and the homeowner's design goals. In this book, we discuss the basics of lighting— why we need it, how it should work in every part of a home, and how to set and achieve your design goals.

HANDCRAFTED LAMPS by Livia McRee
A creative way to decorate your home with personal, custom lighting is to make your own. The craft projects in this book will inspire you to do just that, and produce the perfect light for any place in your home.

Getting the Light Right

We take the meaning of the phrase "let there be light" for granted; illuminating a home in just the right way can be a far from heavenly task. In fact, lighting is such an important feature of our everyday lives that it should be considered when a room is being furnished or, ideally, still in the midst of design and construction. But few of us have the opportunity to build a space from the ground up or even to consider lighting before putting anything else in a room. So we play catch-up, installing lights where we think they should be or where we find we desperately need them.

Regardless of whether we are starting from scratch or backtracking to address inadequacies in the way a space is illuminated, it is critical to develop a lighting plan for each room that addresses both substance and style. The lighting should fulfill the utilitarian objectives of everyday life and at the same time complement a specific decorating style. In addition, a good lighting plan relies on a combination of natural and artificial lighting. It takes into account the available natural light, balancing it with the function, architectural components, and decorative features of a room or space.

left

In this bedroom designed by architect Nico Rensch, individual task lights placed at each side of the bed's headboard supplement the general overhead lighting. A translucent pull-up shade controls glare during the day and offers privacy at night.

START WITH NATURAL LIGHT Thanks to contemporary zoning laws, all the rooms in a home are required to receive some natural light from one or multiple windows, but this does not guarantee that the light is adequate by itself. The quantity and quality of natural light varies greatly from house to house and between individual rooms for several reasons. The way a room—or the house—is oriented, its size and shape, its dimensions and number of windows, and what it is situated next to, all affect its natural lighting.

right

Natural light streaming through a window makes this black-and-white room even more dynamic. Add accent lighting for use when the sun goes down.

Orientation refers to the direction a room faces and is a critical component of lighting. In general, in the northern hemisphere, light from the north is cooler, whiter, and less intense than light from the south, which is why north-facing rooms are favored by artists for their studios (the reverse is true in the southern hemisphere). The cool northern light renders colors more accurately than does the light from the south, which burnishes warm tones. And from the north, intensity is more even and provides steady levels of illumination, as opposed to sharp contrasts of brightness and shadow.

left

During the day, enough natural light filters into the dining area to provide adequate lighting levels. Downlights provide supplement lighting. A fish aquarium built into the wall casts a blue tone.

A room's size and shape also have an impact on lighting. It is easy to assume that a small room needs less lighting than a large one, but that is not always the case, as the direction the room faces, its shape, and the size and positioning of its windows also affect lighting needs. A rectangular room with windows along a short width may need more lighting than a room with windows along its length. A room with a clear view receives more light than one blocked by a building, fence, or tree.

The natural light in a room or space is the starting point for making lighting plans. Although you can't alter the orientation of a home or space, you can compensate for its deficits with artificial lighting.

right
A room can be aesthetically improved with the inclusion of several different artificial lighting sources. Here a floor lamp provides general illumination and also functions as a reading light; a spot highlights the mantel's art; and a task light adds a touch of sparkle.

PLANNING THE LIGHT   A good lighting plan takes thought and consideration, and is ultimately useful only if it is flexible. That means the lighting must be able to conform to changing needs so it can be effective in a variety of situations. It should be able to illuminate all the spaces you plan on using in the room (at different intensities, if necessary) and also be adaptable if you choose to change the layout or function of the room.

To establish the patterns of illumination in a room, first evaluate the activities and tasks that will occur in the space. If it is a large room, this assessment will be more involved, as the space may have to accommodate several activities. Then consider all the wants and needs you have with respect to the lighting. For instance, is the room bright enough from its natural illumination during the day, or would you like it to be brighter—especially in a corner or cranny that natural light doesn't reach? Do you want the room to feel a certain way (warm and cozy or cool and formal, or small and intimate or large and airy) or to accommodate specific tasks (such as homework for the kids and a hobby or craft for you)? In every room, lighting can be used to highlight or accentuate decorative qualities as well as to ensure that your functional needs are met.

left

Provide several lighting options in one room—task lighting for reading, a decorative light for show placed subtly on the coffee table, and a fireplace for ambience.

above

Spot lights hung from a track highlight individual pieces of art and at the same time add general luminescence to the room.

right

Light borrowed from the adjoining room flows through the glassblock wall, while privacy is maintained.

above

Use lighting fixtures to add to the style of the room. Here, candlelight and candle-inspired accent lighting add to the Moroccan design.

The Essentials

When making lighting plans, remember the following:

- If you are starting from scratch and designing and constructing a home, plan for plenty of electrical outlets in each room. This prevents overloading power outlets and eliminates the need for lots of wires or extension cords, which can cause accidents and are unsightly.

- Locate wall switches in easily accessible spots at entrances to each room.

- Wire individual lamps or fixtures in a room to a central switch to make them easier to control.

- Use dimmer switches, which increase the range of effects that can be achieved with the same light and allow you to fine-tune the source of illumination to take advantage of other sources of light.

- Use sensors or timers on lights in areas that require illumination for security, safety, or function, such as high-traffic areas that get very dark at night, entryways, or cabinets and closets.

> **Develop**ing Lighting Plans

Use a basic floor plan to formulate the type of lighting required in every space. Each room needs a combination of three types of lighting: general or background illumination, focal or task lighting, and accent or decorative lighting.

General illumination is also called background or ambient lighting; it is usually the foundation of a lighting plan. It should be used to compensate for lack of natural light during the day, and it should provide uniform illumination throughout a room at night. This uniform illumination is usually provided by ceiling fixtures, but can also be created with a variety of light sources placed around a room that form overlapping pools of light, such as floor and table lamps; up-and downlights; or sidelights, sconces, and spots. While it can be dull and flat or harsh and glaring, overall general illumination should be shadowless, not accentuating anything particular about a space. Instead, it should project a sense of homogeneity in a space and be reassuring and restful. Focal or task lighting is directive—it creates a bright spot that draws our attention, tells us what to look at, or orients us toward an important element or activity center in a space. Task lighting is bright, concentrated, and directed to allow activities to be accomplished with safety and ease, such as preparing food, reading, working, or playing an instrument. However, task lighting also creates shadows around the objects in its field because of the intensity of illumination it throws off. Make sure that the shadows don't fall over the work area; situate the light source in front of or to the side of a person rather than behind him or her. Lamps with long, flexible heads, necks, or arms are usually relied upon to create this type of lighting. If shades that open at the top and bottom (as opposed to highly targeted and closed) are used, focal lighting can also supplement the general illumination in a room.

Accent or decorative lighting is any bright light directed into or onto a specific area for aesthetic rather than utilitarian effect. It is used to enhance or emphasize significant features or furnishings such as architectural elements, shelves, armoires, collections of objects, decorative accessories, and art. Accent lighting can have an unwavering focus to highlight with glamour and drama, or it can wash over a broader area, such as an entire wall or architectural element, for an emphasis that is obvious but more subtle. It can also comprise the accent itself, such as the drama and glitter provided by a chandelier or torchiere. In general, accent lighting should be at least three times brighter than the room's general lighting. Any fixture that can be trained to shine light in one specific direction can be used for this type of lighting. However, if it is too bright, the purpose of bringing nuance and contrast to a space is defeated. For this reason, it is advisable to outfit accent lights with dimmers so they can be adjusted if necessary.

It's critical to utilize all three types of lighting in a comprehensive plan.

> **Decor**ative Tricks

If a room is awkwardly proportioned, too large or small, prosaic and bland, or boring and lackluster, clever lighting can help correct some of these defects.

> **To make a ceiling look higher,** use floor or wall-mounted uplights to throw light up on the ceiling, or conceal lighting behind a cornice or cove mounted high on the walls at the perimeter of the room.

> **To make a ceiling look lower** and give a room an intimate demeanor, keep light away from the ceiling by placing wall lights fairly low and using shades or pendant fixtures with closed tops that won't throw any light back on the ceiling. Also, draw attention to items placed at a low level, such as pictures or wall hangings positioned low on the wall or groups of accessories placed on low surfaces, by lighting them from above or with a lamp standing on the surface.

> **To make a long, narrow space seem wider,** focus attention on a feature on one of the end walls, such as a window with an elegant treatment or an interesting piece of art, by highlighting the wall with a spotlight, and wash the other walls in the room with an even but less intense light.

> **To make a space seem larger,** wash opposite walls with light to make them seem farther apart. Alternatively, combine lighting with mirrors and reflective surfaces to increase the illusion of space. Use recessed cans or track-mounted spots, pointed downward, above a large wall-mounted mirror, or position lamps to be reflected in a mirror.

> **Surfaces with sheen—such as glass,** metal, tile, and gloss paint—reflect, heighten, and amplify light, while matte or textured surfaces absorb light rather than reflect it. The darker in tone, the more light is absorbed. Use matte surfaces in an overly bright room to cut the glare from too much sunlight.

> **To make a space more intimate,** use several table lamps to create a cozy glow. Make sure they have shades in colors that are warm rather than cool, such as alabaster, pearl, parchment, or ivory instead of white, so the light they cast has a mellow tone.

> **Window treatments have a huge impact** on light and can be used to manipulate the mood in a room. Translucent drapes, which filter natural light and produce a diffused effect, come in many weights and should be chosen with regard to this property. Slatted blinds offer optimum light control, ranging from total transparency to full screening, and also create dramatic patterns of light and shade when they are opened at various angles.

> **If there are several light sources** in a room and you are unhappy with the overall effect, change the bulbs, which may be too bright or too dim for their surroundings. If a fixture or lamp emits too much glare, replace its standard bulb with a reflector bulb.

> **To change the lighting in a room,** change the shades in a fixture. Translucent materials allow more light into a room, which makes it seem brighter, while darker shades can glow softly when the light is turned on, lending drama to the space.

> **To imbue a space with color,** which can create a specific mood, bathe a white wall with a washer fitted with a colored filter. Before doing so, research the psychological impact of the intended color. Yellow is a cheery hue that promotes feelings of well-being and increases efficiency. Blue is a soothing hue that calms nerves and induces sleep. Red is stimulating and dynamic and enhances the action wherever it is used. Green is a harmonious color that reminds us of nature and can be warm, refreshing, and earthy. Purple can be calming or exciting, depending on whether it leans more toward blue or red.

Color and Light

- Pale walls reflect natural light and spread it around the room. Darker, intensely colored hues absorb light and give a space a heightened sense of enclosure.

- White ceilings act as a huge reflector for natural and artificial light.

- Colors from the warm end of the spectrum—such as yellows and creams—can take the chilliness out of north light, while cooler colors—such as blues and greens—can be dazzling in direct sunlight.

- Every color changes shade under different lighting conditions throughout the day, which affects and reflects everything else in the room. Keep this in mind when choosing the hues of walls, drapes, and large pieces of furniture.

_above

By day, the kitchen is illuminated by natural light flowing in through the large skylight; by night, pendant lights provide ambient lighting.

> **The** Psychology of **Lighting**

Lighting has a dramatic psychological effect on how we perceive a space and affects how we feel when we use it.

For instance, sunny rooms that get lots of bright light are welcoming, warm, and cheery, making us feel good, while rooms that receive indirect light can be dull, lifeless, and cold, leaving us depressed. People also feel alert, energetic, and positive on a sunny day, which causes bright highlights and crisp shadows in a room, and the opposite on a dull, drab day, when there is no contrast and the environment is stagnant, boring, and uninspiring. The difference between these two days can be chalked up to variations in the quality of light; the proper illumination can provide contrasts in a room that emulate the attributes of a sunny day.

Environmental psychologists evaluate the stimuli that must be processed in a room by those using it, which is relevant to how a space should be lit. Rooms that are crowded, asymmetrical, disorderly, unconventional, or unfamiliar have a lot of stimuli and are considered high load, while rooms that are straightforward, symmetrical, conventional, familiar, and organized are less arousing and are considered low load. Tasks are gauged in the same way. Doing something demanding, such as reading a challenging book or writing a complex essay, is a high-load task, while tasks that are simple or routine, such as paying bills or cleaning the house, are low-load tasks. Lower-load tasks require higher-load settings for optimum performance, and vice versa. Lighting can be used to increase or decrease stimulation by creating an emotional setting in a room that affects the performance of tasks. In fact, the proportions of the three types of lighting—background illumination, task lighting, and accent lighting—determine the emotional content of a room.

A space lighted with a large proportion of background illumination, evenly diffused, and a small amount of focused task lighting or decorative accent lighting, is a low-contrast, low-stimulation space that is behaviorally neutral, as it is minimally stimulating. This type of environment is ideal for performing visual tasks such as reading or working. But too much diffuse light produces a boring, shadowless environment, which can evoke the type of bland psychological reaction experienced on a cloudy day.

A space lighted with a small amount of diffuse light and a larger amount of focal light is a high-contrast environment featuring strong patterns of light and shade. This type of lighting plan increases stimulation and is intended to evoke specific moods and emotions. A room lit in this manner can also dominate the people in it; the contrast produces visual direction and focus by directing their attention and holding their interest.

Overall, people need lower-loaded settings for difficult, complex tasks or to feel contented, comfortable, and relaxed, and higher-loaded spaces for casual, pleasant activities or socializing, as a high degree of contrast encourages participation and stimulates enjoyment.

Light strings, so essential for a festive mood during the holidays, can be fun any time of the year and may be just the thing to liven up a theme party. Medium-weight papers will hold their shape when folded, but are still thin enough to let light shine through. Personalize your shades for any kind of party using specialty punches like the sun we used, and different paper edgers. You can also try lining your shades with white or colored vellum for a softer light through the punches.

Punched Paper Light String

Materials

- Light string
- Star and sun paper punches
- One large sheet (about 19 inch x 25 inch [48 cm x 64 cm]) of medium-weight art paper, such as pastel paper, of each of the following colors: red, yellow, and orange
- Double-stick tape
- Ruler
- Pencil
- Scissors or a craft knife
- Deckle edge paper edgers

TIP Make a sample lantern and see if you like the size in relation to your light string. Some light strings may have close-together lights that will look better with smaller shades; simply reduce the panel height and width proportionately to make a smaller shade.

1. Measure and cut paper for lanterns.

From one of the colors, using the ruler and pencil, measure and mark a 6-inch-tall by 12 1/2-inch-wide (15 cm x 32 cm) piece of paper; repeat three times for a total of four panels. From the other two colors, measure and cut three panels each.

2. Fold lanterns.

Using the ruler, measure up 2 1/2 inches (6 cm) from one of the long edges, and draw a line from short edge to short edge; then mark a line 1 3/4 inch (4 cm) up from the same edge. Place the ruler on the 2 1/2 inch (6 cm) mark and fold the paper up along it, then remove the ruler and make a sharper fold. Do not fold along the other line. Repeat for all the panels of paper.

Next, fold each panel into fifths, using the ruler and technique described above; each section should be 2 1/2 inch wide (6 cm). Make each fold in the same direction to form the shade.

Make flaps for the shade by cutting along the creases that have divided the shade into fifths, starting at the longer edge of the paper where you measured up 2 1/2 inch (6 cm) to begin with, and stopping once you get to the horizontal fold. Using the previously marked line 1 3/4 inch (4 cm) from the edge of the paper that you didn't fold as a guide, trim the first, third, and fifth flaps; they should measure 3/4 inch (2 cm).

Punch a random design in the second, third, and fourth sections at this point.

3. Fold lanterns and finish the shades.

Bring the first and last panels together and align them; secure with double-stick tape rather than glue to avoid warping the shades.

Punch the fourth side of the shades. Then, gently flatten each shade and trim the edge with the paper edgers.

4. Attach shades to the light string.

Place a light in between the short flaps of one of the shades, then fold the two longer flaps over the string and each other; secure with double-stick tape.

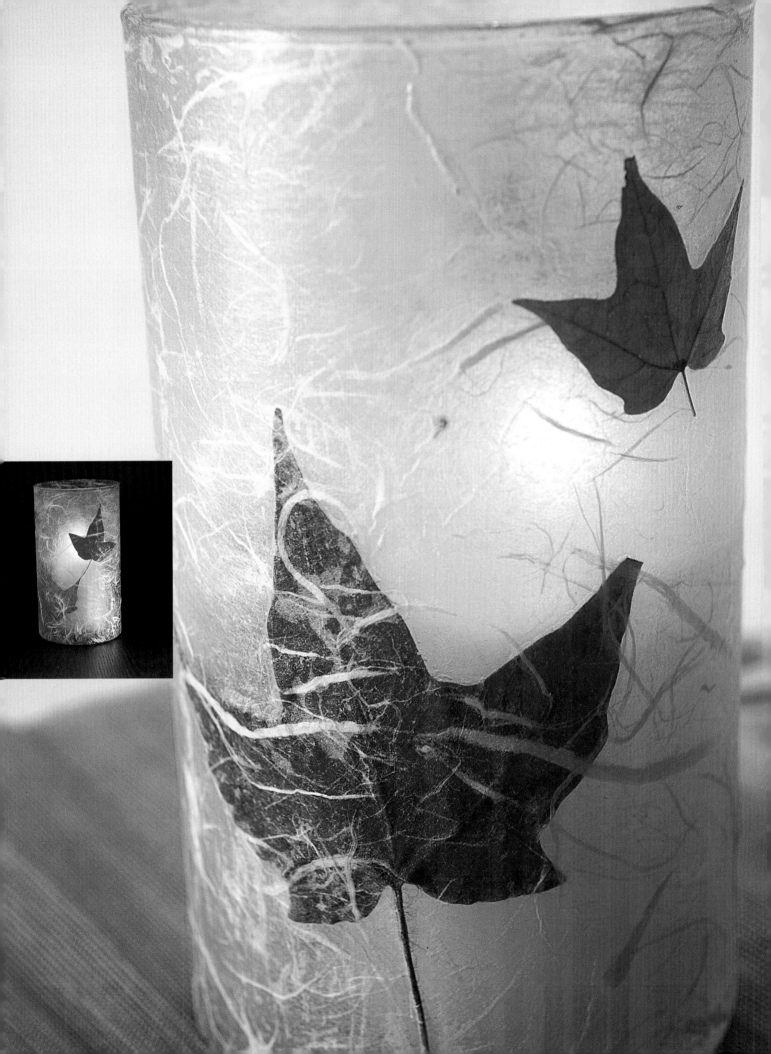

Handmade paper, available in a variety of styles and colors to match your taste and home, is even more appealing when combined with pressed leaves or flowers. This hurricane, which can be assembled in under an hour, can be easily adjusted to fit the decor of any room. Be sure to test the translucency of the paper you select by holding it up to the light. You can apply this technique to any glass container you want, or use bits of torn paper rather than a whole piece to cover the container for a mosaic or collage feel.

Leaf and Paper Hurricane

Materials

- Glass container, such as a cylindrical or square vase, or a hurricane shade
- Several pressed leaves or flowers
- Thin, handmade paper
- Ruler
- Pencil
- All-in-one découpage glue, sealer, and finish such as Aleene's gloss finish
- Foam applicator brush

TIP You can press your own leaves in a telephone book until dry, approximately one or two weeks, depending on humidity. Many craft stores offer skeletonized leaves, which have been treated so that only the veins and stem remain, resulting in a subtle, ethereal effect.

For easier lighting and the best effect, choose a candle that is about half as tall as the container.

1. Measure and cut the paper.

Place the glass container on the paper, and wrap the paper around it. As you wrap, mark a line for cutting with a pencil along the edges of the container at the top and bottom. Then unwrap the container, and add 1/4 inch (.5 cm) to the length for the seam. Next, tear the paper by laying a ruler along your marks and gently pulling the paper toward you. The irregular edges will produce a softer look and will help disguise the seam.

2. Arrange the pressed leaves or flowers.

Clean and dry the container. Apply a thin, even coat of glue on the vase where you want the leaves or flowers to be, using sponge applicator. Arrange them as desired, and gently smooth the leaves outward from the stem with the sponge applicator. Be sure there are no wrinkles or air bubbles in the leaves. For brittle but thick foliage, gently coat the back with glue first, which will make it more pliable.

3. Apply the paper to the vase.

Gently apply a thin, even coat of glue to the entire container, including the leaves or flowers. Position one edge of the paper on the container where you want the seam to be and begin wrapping it on smoothly and slowly. Smooth out wrinkles and bubbles as you go along. Finally, apply another thin, even coat of glue over the paper and let it dry overnight before using.

Lighting Living Areas

In every room in a home, there is a constant give and take between natural and artificial light. This ever-changing dynamic unfolds over the course of an entire day and night. Thus, the features of a room—such as its dimensions and architectural details; the size, type, and number of its windows; its orientation to the world outside; and the way it is used throughout the day—are all key elements that must be considered from sunup to sundown when lighting that room. Ultimately, it is important to pay as much attention to the effect sunlight has in that room as it is to consider the impact that artificial lights have in the same space.

Few of us realize all this when setting out to light our living areas. Traditionally, a standard-issue lighting plan has evolved for each of the public spaces in our homes. The living room contains several seating or activity areas, as it is a multipurpose space for most of us, and these are generally lit with floor lamps. An additional form of general illumination is usually present overhead, either in the form of tracks, a central ceiling fixture, or recessed cans. In dining areas, the dramatic pendant fixture or chandelier reigns supreme. In bedrooms, the bed or an easy chair off to one side is usually flanked with reading lamps, while the entire room is illuminated with a ceiling fixture. While this standard approach is tempting, it is not the most effective way to light a room.

left
A simple table lamp supplements a living room's ambient lighting while also serving a decorative purpose.

In a perfect world, we would be able to address the lighting needs of a room before it is furnished, or even while its blueprints are still in on the drawing board, but in reality, it is impossible for most of us to start from scratch. We move in and out of residences with more frequency than ever before, sometimes physically altering them by installing the sort of lighting we need when we get there. Often, we address immediate lighting needs with store-bought fixtures, yet carefully planned alterations to lighting at any stage can radically improve a room from both a functional and an aesthetic perspective. When those rooms are the ones we use most of the time, these changes become even more meaningful.

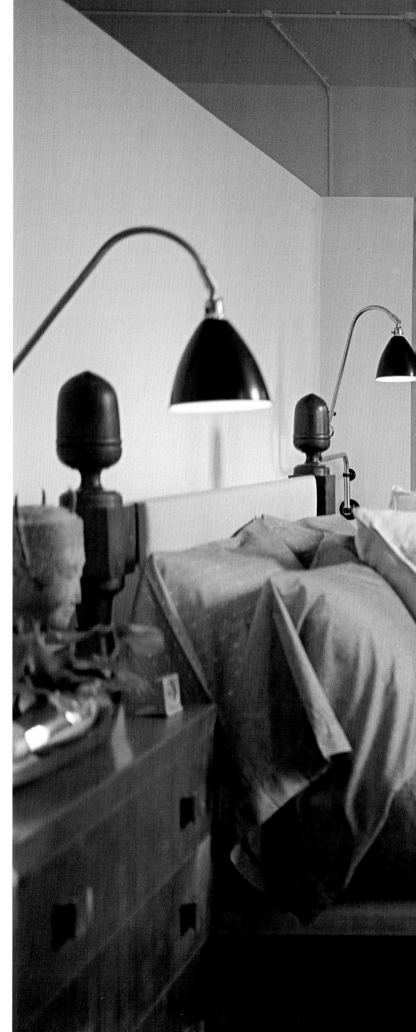

right
When the daylight fads, a hierarchy of artificial lighting takes over—ranging from the two clip-on task lights over the bed pillows to the focused desk lamp to the long-armed flexible floor lamp to the romantic glow of the fire.

LIGHTING PUBLIC LIVING SPACES Today, we often spend the bulk of our time in open, airy rooms used for a variety of activities, such as relaxing, watching television, reading, listening to music, and entertaining friends, or even working on hobbies, homework, or special projects. We refer to these spaces with an assortment of terms, such as great room, family room, den, or even the plain, old-fashioned-term living room, but in essence, they are hardworking, multipurpose spaces that have to accommodate a range of disparate endeavors.

Just as lighting draws attention to actors on a stage, it must highlight and accommodate the various parts of the set in these sorts of rooms. At the same time, the lighting must play up the room's best assets. It is necessary to focus adequate amounts of the right kind of light on various furniture groupings and, at the same time, complement and enhance the decor of the room. Reconciling these needs requires a combination of the three types of lighting (namely, general, task, and accent lighting (see Developing Lighting Plans in Chapter 1).

right
By day, natural light heightens the color of this traditional drawing room. At night, the yellow walls, lit by an overhead fixture and two table lamps, mellow and a burning fire brings a gentleness to the room.

Modern-day chandeliers come in all shapes and forms—from unusual to classic, as seen here.

In dining areas, or those portions of multipurpose spaces that are used for dining, the chandelier is still the fixture of choice—for good reason. Not only is it a suitable and often even superior source of light, it is an exceptional decorative asset that can go a long way toward expressing the specific style of a room. A frothy crystal version adds elegance and grandeur to a space, while a rustic wrought-iron fixture can convey intimacy and warmth. Today, pendant fixtures are available in a broad range of styles and are used with the same frequency as the chandelier.

It is important to keep several considerations in mind when selecting a hanging fixture for a dining area. The fixture should complement the size and shape of the dining table as well as the decor of the room. Before electricity was commonplace, such lamps were fitted with candles and hung high over a table to eliminate the danger of fire, but now, a hanging fixture should be suspended about 30 inches (762 millimeters) above the table. If the fixture has an open shade and bare bulbs, it should be suspended as high as necessary to avoid a harsh glare in diners' eyes, or a bulb with a silvered crown should be used to reduce glare. Also, the diameter of a hanging light should be at least a foot shorter than the table below it, and when the ceiling height is over 8 feet (2.4 meters), balance the space by raising the fixture 3 inches (76.2 millimeters) for every additional foot of ceiling space. Tame dramatic but too-dazzling pendant fixtures and chandeliers with a dimmer control. Finally, remember that no rule dictates the use of a hanging fixture. Recessed cans or tracks equipped with downlights can also be used to adequately light a dining table, which can be accented with flickering candles to change the ambience or mood.

—**left**
Modern-day chandeliers come in unusual shapes and forms.

LIGHTING PRIVATE LIVING SPACE Gone are the days when bedrooms were meant solely for sleep. Today, they have become multi-purpose rooms and/or private sanctuaries for many members of the family. Sometimes, bedrooms are even entire suites that span several rooms for those lucky enough to have such space at their disposal. All this makes the lighting requirements in our bedrooms as varied and complex as any other living space in the home.

Just consider all the activities that take place in this room, from reading, writing and watching television to rummaging around for what to wear and dressing. All these activities, save sleeping, requires a specific source of light, and if two people are sharing the space, it calls for sources of light that are flexible enough to be adjusted to two activities at once. For instance, if one person wants to read or must get dressed while the other is sleeping, the lighting in the room should be deftly planned to accommodate such subtle or specialized adjustments. All this reinforces the idea that a bedroom needs both general illumination and task lighting. Often, accent lighting also comes into play as a decorative element.

right
The lighting of a bedroom dresser takes a more personal touch with lamps that provide ambient lighting but also a touch of whimsy.

So where to start? In fact, the traditional ceiling hung or mounted fixture is entirely inappropriate in the bedroom for many reasons. In this particular room, points of activity tend to be situated around the perimeter of the room rather than anchored in the center. Beds are usually positioned against walls rather than in the center of a bedroom; a reading area is often incorporated into the bed set-up with task lighting or sequestered in intimate corners with the aid of easy chairs and lamps; and dressers or bureaus are also placed against walls. All this necessitates lighting sources that are tied to the layout of the room and the needs of its users. Also, if there are translucent shades or drapes in a bedroom, ceiling mounted fixtures can unwittingly silhouette intimate moments.

Ultimately, because of the nature of the bedroom and its most important activity, namely sleeping, this is one area of the home where flexibility is paramount. So it pays to increase the variability of light sources in the space by outfitting them all with dimmer switches. To read in bed, besides typical bedside lamps, swing-arm fixtures can be mounted to the wall, or for the most maneuverability, affixed directly to the bedpost with screw-tightened vises so they can ride up and down for the best positioning. Or consider spotlights or reading lamps on clips that can be used on a headboard or bedframe. And finally, don't forget to have two main light switches in the room: one by the door and one by the bed, since there is nothing more irritating than getting out of bed when you're almost asleep to turn off the lights.

Fixture Formulas

To get the most out of a fixture, position it optimally with regard to its purpose. For instance, a pendant fixture intended to provide ambient or background illumination casts a broader path of light the closer it is to the ceiling, but if it is meant to provide task lighting, it must be set lower. Here are guidelines to follow:

- Lamps used for reading: Both floor lamps and table lamps next to chairs have the same requirements. The bottom of the shade should be about 40 inches (1,101.6 millimeters) to 42 inches (1,066.8 millimeters) above the floor, which is slightly below eye level for a seated reader of average height. Lamps behind chairs should be taller; the distance from the floor to the base of the shade should be a minimum of 47 inches (1,193.8 millimeters). The lamp should be placed approximately 10 inches (254 millimeters) behind the shoulder of the reader. Next to beds, the base of a lampshade should be 20 inches (508 millimeters) above the pillow.

- Task lighting: The height of fixtures and lamps varies by task, but rules of thumb are available. A pendant fixture should be about 30 inches (762 millimeters) above the top of a dining table, but if a room is over 8 feet tall (2.4 meters tall), add 3 inches per foot above 8 feet to this figure. Light sources used for light-intensive tasks, such as working, writing, sewing, drawing, or working with tools, should be 14 inches to 15 inches (355.6 millimeters to 381 millimeters) above the center of the work, which should be positioned 10 inches to 14 inches (254 millimeters to 355.6 millimeters) in front of the worker.

- Wall-mounted fixtures: Heights vary and depend on the way the fixture is designed and the height and size of the wall, but, in general, they should be placed above eye level. They should also be fairly flat, not protruding more than 4 inches (101.6 millimeters) from a wall, unless they are placed well above the level where they can be bumped by heads (which is at least 80 inches {2,032 millimeters} above the floor).

- Track lighting: The beauty of this type of lighting is its flexibility; the fixtures can be moved along the track to where they are needed and positioned at any angle to become a downlight, spotlight, or accent light for art and collectibles. To accent an item on a wall, position the fixtures at a 30-degree angle.

right
The light from two column torchieres is doubled as it bounces off full height mirrors in the library-music room. Swirly decorative torches holding candles flank the fireplace and add a modern romantic touch.

right
A cascading chandelier highlights the dining area, while a
funky floor lamp provides
reading light in the sitting
room.

> **Plan**ning Points

Consider the following points when planning a lighting scheme for a room:

> Activities: What will be happening in the room, and where will each specific action take place?

> Highlights: What features in the room need emphasis? Stunning architectural elements, such as magnificent mantels or intricate ceilings, should be bathed in light to show them off.

> Deficits: What needs to be concealed? Faults, such as ugly architectural ornamentation or badly plastered walls, should not be emphasized with light.

> Ambience: What is the desired mood for the room? Dramatic or soothing? Businesslike or cheery and bright? The choice dictates the intensity and positioning of the light sources in the room.

> Balance: Should the entire room be brightly lit, or would pockets of brightness and shadow be desirable? If the latter, make sure these pockets balance each other and do not detract from the room's functionality.

> Flexibility: To accommodate different activities in the room, make sure the level of general illumination is adequate and employ free-standing fixtures that can be moved at will.

> Variety: A lighting plan that uses all the same fixtures or lamps can be both boring and inadequate, as it is unlikely that one type of item can satisfy all requirements. Use a diversity of light sources in a room, and choose attractive and creative options.

> Decorative style: Make sure the light sources complement or match the decorative style of a room. If a light source is particularly stunning in itself, such as an intricate chandelier or an artist-made lamp, it can even be the focal point of the space.

> **Find**ing the Right Fixtures

A good lighting plan is only the beginning of the process to follow when lighting a home. After identifying the lighting requirements of a room, the next step is to fulfill them. This often calls for employing all three types of lighting in a space (see Developing Lighting Plans in Chapter 1), and it also takes careful planning. At times, it may even demand a bit of experimentation.

Once you know the types of illumination needed in a room, it is possible to select fixtures and lamps. While it isn't necessary to master the complex calculations of the professional designer, a working familiarity with the different types of lights and fixtures is helpful in making an informed choice. Consider the quality of light each fixture gives off, its aesthetic appearance, how sturdy it is, and the type of light it emits (see sidebar, "Types of Light"). Also, keep in mind that no light can serve every purpose or meet every need.

Lighting fixtures come in every form, shape, size, and style imaginable, and are made of myriad materials. In fact, the possibilities are endless, and there is no such thing as one right choice. However, all these options boil down to three general categories and, when lighting a space, it is usually necessary to choose fixtures from each of these groups to create schemes that balance function with aesthetics.

> Freestanding Fixtures

This type of lighting can range from soaring torchieres to squat table lamps. Sometimes freestanding fixtures are stunning works of art in their own right. Basically, however, they can be used to address every lighting need, depending on the style and properties of the particular lamp. Torchieres are uplights, which can be used to create ambient light, while floor and table lamps usually sport shades that suit them for focal or task lighting. Freestanding fixtures have the flexibility to be moved wherever you need them and don't have complex installation requirements; you merely plug them in. There are two basic types of freestanding lamps: uplights and lamps with shades. (The term lamp is usually used to refer to freestanding fixtures but, technically, a lamp is the part of a bulb that emits light.)

Uplights work best in rooms with high ceilings, where they make an ideal source for general illumination. Uplights can use the ceiling as a giant reflector to create a softly diffused ambient light in a space, provided the ceiling is painted white or a light color (dark colors absorb light). The height of an uplight determines the intensity and quality of the light it yields as it reflects off the ceiling; those close to the ceiling produce a concentrated light, while those that are lower produce light that is softly diffused. In general, the greater the distance between the uplight and the ceiling, the greater the area illuminated.

Lamps usually consist of a base or stand that supports a socket, the bulb, and a shade. The power cord is concealed by the base, and the shape and substance of the shade determine the quality and quantity of light the lamp emits. The beauty and versatility of the lamp lie in this simple design, for the base and shade can be made of myriad materials, take virtually any form, and be rendered in any decorative style. Also, the shade can direct the light of the bulb in any direction, though most focus the light downward to create a softly diffused pool of illumination. The wider the shade, the broader the pool of illumination.

above
A pyramid-shaped floor lamp with tapered linen shade.

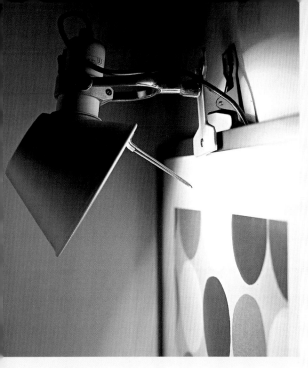

> Ceiling Fixtures

Illumination from above is one of the most basic ways to light a room, and depending on how they are installed and the intensity of the bulbs, ceiling fixtures can provide background or general illumination in a room as well as task and accent lighting. There are many ways to mount fixtures on the ceiling, and basic ceiling fixtures are usually made of translucent glass, sport simple curved or geometric shapes, and are attached to, rather than hung from, the ceiling. Either one or several fixtures are used to provide general illumination in a room. They are also usually used in high-traffic areas, such as hallways, kitchens, and bathrooms, especially when ceiling heights are low. Five types of ceiling fixtures offer specialized and directed lighting options: pendants, chandeliers, downlights, spotlights, and tracks.

Pendants hang from the ceiling and can be used to provide general or task illumination, depending on how they are finished off. The bulb can be covered with translucent globes for diffused ambient illumination that radiates throughout the room, or trained in a specific direction with shaped shades for task lighting. Small pendants can be used at various points in a room, or even grouped together, for more illumination and for their combined decorative effect. They can be all the same or related in terms of color or styling.

Chandeliers, which are pendants with branches that hold bulbs (or,

in some cases, candles) as well as decorative ornamentation, are the most dramatic type of ceiling fixture. The quality of light produced varies with the number, type, and strength of the bulbs, the style of the fixture (some have many branches or sport individual shades over each bulb), and whether or not the fixture is equipped with a dimmer. A chandelier is the focal point of the space or room it occupies and must be chosen accordingly.

Downlights are minimal ceiling fixtures that are inconspicuously mounted on or in the ceiling to cast light directly down on a surface. They can be surface mounted, recessed, partially recessed, and used to provide all three types of lighting, but are particularly effective at focal illumination. The beam of light downlights

emit can vary in width, depending on the shape of the fixture and type of bulb that is used; recessed lighting produces the narrowest beam. Downlights are often used over work counters for task lighting, or to accent or emphasize a particular area or feature of a room. While they are ideal for low ceilings where pendant lights would be unsuitable, fully recessed lighting requires a half-foot of space above the ceiling to accommodate the fixture and provide proper ventilation.

Spotlights, like downlights, are minimal fixtures that are available in a wide variety of shapes and styles, but they are extremely adjustable and are used to focus on, or emphasize, specific spots in a space. They can be installed or mounted on walls, floors, ceilings, or standing polls, set in

tracks, or used individually with clips for greatest flexibility. They are best for accent lighting, as the bulb and fixture are designed to work together to give a precise, controlled beam of light that can be tilted, swiveled, or angled to wherever it is needed. However, in large, open spaces, spotlights can be used for general illumination by positioning them to create overlapping pools of light. This technique produces ambient lighting that is a bit more interesting than the uniform level of illumination provided by basic ceiling and pendant fixtures.

Track lighting is often referred to as its own type of lighting, but it is actually a combination of spotlights and downlights installed in a flexible arrangement on a ceiling-mounted track. Track lighting can be used for general, task, and accent lighting because it is possible to swivel, rotate, or point the individual fixtures in any direction; often, the same track incorporates all three types of illumination. Track lighting comes in two forms. Some tracks are fixed and come with specific light fixtures already attached, while others are merely free-form power lines that can be used to anchor a number of types of fixtures along its length.

> Wall-Mounted Fixtures

Like ceilings, walls provide a functional and accessible surface for mounting fixtures, and, like lamps, fixtures designed to be wall-mounted come in a tremendous scope of styles ranging from period or traditional versions to cutting-edge designs. Some emit an all-round glow, such as those encased in glass globes or translucent lanterns, while others train lights either up, down, or at a specific spot due to the way they are shaped. In most cases, these fixtures are in harmony with the walls they grace, using the surface as a reflective plane to create a subtle source of ambient light in a room. Unlike ceiling lights, which usually generate a steady level of light in a room, wall lights can be used to create pockets of light or dramatic ebbs and flows in a room.

Sconces are a common form of wall-mounted lighting; they were once ornamental wall brackets meant to hold candles, but they hold electric bulbs today. Some sconce styles take candle-shaped bulbs. When such bulbs are left exposed, opt for lower wattage to avoid glare. Sconces are often used in pairs and, depending on their design, become a focal point when employed in this way. They can also instill a sense of symmetry and drama in a setting when used to surround a piece of furniture, such as a bureau or bed, or an architectural element, such as a mantel or a doorway.

Wall washers, another form of wall-mounted lighting, are usually installed at the top of a wall, pointing down, to bathe the walls with a steady, uniform, and soft level of illumination. With these fixtures, the results are most effective when the wall is smooth and painted a light color to maximize its reflective properties.

above
Add this sconce and cord cover without drilling a hole in your wall.

> **Balanc**ing Acts

The lighting fixtures that are used should be balanced in terms of the furnishings that surround them, whether sofas and tables or other sources of light. They should also match or complement the style of the decor in a home. Keep these points in mind:

> **Lighting fixtures** placed near each other should be balanced in terms of weight and the amount of light they emit. For instance, a very tall lamp placed next to a short lamp overwhelms it, and the light emitted by both lamps is uneven and perhaps even conflicting. The same is true of wall sconces, which should be kept consistent in style and light emission throughout a room, especially if they bracket an architectural or decorative element such as a fireplace, mirror, bureau, or doorway.

> **Relate the scale** of a fixture or lamp to the scale of the furnishings that surround it rather than the scale of the room. If a small space is filled with a chunky sofa, pair it with a chunky table lamp with a wide shade. Hang a husky pendant lamp over a burly, masculine dining table; pair a curvy, delicate sofa with a like-minded table lamp or floor lamp; use weighty sconces to surround a massive stone hearth.

> **Stay within the parameters** of a room's decor. If the style has a country demeanor, use lamps made of down-home materials such as earthenware crocks and mason jars. In a formal milieu, stick to prim and proper fixtures in elegant materials with graceful shades. In a period room, such as one with art deco or modernist furnishings, use vintage or reproduction pieces that approximate or imitate the style.

> **When a fixture** is paired with an important piece of furniture in a room, it should relate to the piece. Complement a contemporary glass or marble dining table with a fixture that echoes its shape or the materials used to fabricate it. For instance, pair a round glass table with a round glass pendant, or repeat a geometric motif in the fixture over a square marble-topped table. Surround a lacquered oriental armoire with lantern-style sconces.

> **Be sure to bring** decorative details and incidental furnishings into play when possible. Match the color or pattern of textiles in a room with shades in the same colors or motifs, or use the same trim from pillows or drapes to edge shades. Surround a gilt-framed mirror with gilt sconces, reflect wrought-iron curtain rods with a wrought-iron chandelier, or get a fixture with a stainless-steel shade to complement stainless-steel kitchen counters.

above
Matching geometric-shaped table lamps flank the sofa and reflect the décor's stylistic leanings. The decorative lamp atop the cabinet adds delight as well as light.

The simple appeal of a Japanese-styled lantern can add a clean, modern look to a bedroom or dining table. With simple supplies and elegant, authentic Japanese paper, you can create a lantern of your own without a hammer and nails. Craft and hobby wood, such as balsa or basswood, works best because it is very soft and easy to cut. For a more traditional look, divide the panels into evenly-spaced multiple panes.

Japanese Table Lantern

Materials

- All-purpose clear-drying glue
- Balsa or basswood strips: four 1/2-inch-wide (1 cm) strips, and four 1/4-inch-wide (.5 cm) strips (these are available at hobby and craft stores and come in approximately 24-inch-long (61 cm) strips)
- Thin piece of plywood, no less than 1/8 inch (.3 cm) thick but no more than 1/4 inch (.5cm) (small pieces are also are available at hobby and craft stores)
- Thin, handmade rice paper or something similar
- White cardstock that matches the rice paper
- Light-colored wood stain and finish in one
- Sponge brush applicator
- Ruler
- Pencil
- Craft knife
- Hobby saw attachment for craft knives
- Gridded cutting mat
- Fine sandpaper

TIP Use a cotton swab to apply small amounts of glue precisely and evenly. Use sandpaper to make the lantern's legs level and steady.

1. Measure and cut the wood.

From the 1/2-inch-wide (1 cm) strips, cut eight 12-inch-long (30 cm) strips using the hobby saw. From the 1/4-inch- wide (.5 cm) strips, cut eight 5 inch (13 cm) strips, eight 2 3/8 inch (6 cm) strips, and four 9 1/2 inch (24 cm) strips using the hobby saw. Cut the plywood to be 6 inch (15 cm) square using the hobby saw.

2. Measure and cut the paper.

Using a craft knife and a gridded cutting mat, cut four pieces of paper 6 inches wide by 10 inches tall (15 cm x 25 cm). Make sure that each piece is exactly the same size, and properly squared up.

To make reinforcements for the corners of the lantern, cut four 1/2 inch wide by 9 7/8 inches (1 cm x 25 cm) long strips from the white cardstock. Then, measure in 1/4 inch (.5 cm) and lightly score them with the craft knife down the length; fold the paper to make a 90 degree angle. From the rice paper, also cut four 1/2-inch-wide by 9 7/8-inch-long (1 cm x 25 cm) strips and glue them into the corners; they do not need to be scored because the paper is very thin and easily folded.

Also make reinforcements for the inside bottom of the lantern where the panels meet the base of the lantern by repeating the above technique; make four more reinforcements 5 7/8 inch long (15 cm).

3. Assemble the four panels of the lantern.

To assemble each panel, lay a piece of paper on the gridded mat and glue a 1/2 inch-wide (1 cm) piece of wood, flush with the edge of the paper, to both long sides with 1 inch (3 cm) extending above and below. Use the mat's markings to be sure the wood is properly aligned with the paper.

Next, glue a 5 inch (13 cm) strip to the top and bottom shorter edges of the panel, flush with the edge of the paper. Be sure to add a dab of glue to each end of the strips as well. Then glue a 9 1/2 inch (24 cm) strip to the center of the panel. Finally, add a 2 3/8 inch (6 cm) strip down 3 inches from the top strip on the left side of the middle strip, and another 2 3/8 inch (6 cm) strip up 3 inches from the bottom strip on the right side of the middle strip.

4. Assemble the lantern.

Once all the panels have dried, glue each to one side of the 6 inch (15 cm) square base, aligning the bottom edge of the base with the bottom edge of the paper; use glue only at this point. Wait for the glue to set, but not dry completely before going on to the adjacent side. Glue the reinforcements in place as you create the side and bottom corners.

Papyrus, with it's tartan-like weave, adds visual interest to a simple shade and sheds a warm, soft glow. A papyrus shade paired with a reclaimed treasure from your attic or a yard sale makes a unique and personal accent for any room. Bottles are easiest to convert to a lamp base, but with a little extra effort, you can transform just about anything. Once you've assembled one lamp, you'll see how easy it is to make one from anything, such as a vase, pitcher, or ceramic crock.

Papyrus Shade Lamp

Materials
- A container to convert to the lamp bottom, such as the antique seltzer bottle seen here
- Self-adhesive lamp shade in a size appropriate for the base
- A large sheet of papyrus
- One package of 1/4 inch (.5 cm) double fold bias binding, off-white
- Wiring and lamp hardware kit for bottles (available at hardware stores)
- Fabric glue
- Scissors

TIP When selecting paper for a shade, always check what it will look like when backed by a light bulb and a layer of thick white paper, to simulate the shade backing. Thicker papers may not diffuse the light enough, and the patterns of very thin papers may be washed out. To make the pattern of a thinner paper show up better, try gluing it to a piece of thick white paper with spray adhesive, then attaching it to the shade.

1. Cut the paper for the shade.
Tape the paper to your work surface, and trace the shape of the lampshade wrapper on the paper with a pencil. Cut the paper along this line.

2. Attach the paper to the shade.
Starting at the seam, wrap the shade with the paper. Use the fabric glue to secure the seam.

3. Add the trim.
Before cutting the trim, tape one end to the top of the shade and wrap it around to see how much you will need. Cut at this point, then repeat to measure the amount needed for the bottom of the shade. Using fabric glue, secure the trim to the edges of the shade, inside and out.

4. Assemble the lamp hardware and fit it to the base.
Many hardware stores have kits for wiring bottles that contain several sizes of gaskets for fitting the lamp hardware into the bottle's opening, and the cords are designed to come straight out at the socket rather than down into the bottle. This allows you to preserve the look of the transparent container.

Following the kit's instructions, attach the cord to the socket. Then attach the "harp" to the socket. The harp is what the lampshade will attach to; a taller harp will show more of the base, and a shorter harp will conceal some of it. Using a finial, and washers for a tight fit (if necessary), attach the shade.

If you choose a bottom that has a larger opening than a bottle, such as an antique vase, the assembly will be virtually the same, but you will need to get the proper hardware from an electrical or lighting supply store, or from a mail-order source.

Lighting Hardworking Areas

Bright, concentrated light is an essential ingredient in the hardworking areas of a home. It keeps us alert, focused, and even safe at what can sometimes be hazardous tasks, such as cutting vegetables with sharp knives. Task lighting is therefore the choice for areas where concentrated activities repeatedly occur—in the home office or study, the kitchen, the bathroom.

Task lighting, be it a table lamp or a focused, recessed light, casts a bright directional beam on a defined area, yet achieving visual comfort in hardworking areas demands accurate placement of the light as well as merely selecting the fixture. The light should be positioned to avoid excessive glare, reflectance, and shadows. What use is task lighting if you are working in your own shadow or if the light bounces off a polished surface nearby and the excessive glare bothers you?

left
With lighting fixtures strategically placed—the chandelier, the track lighting, and the downlights over the sink—make this kitchen bright and functional.

To understand this concept, think of the structure of the human eye. The pupil automatically enlarges or contracts to admit more or less light. Alternating excessive and limited light levels tires the eye muscles. Therefore, bright task lighting in an otherwise dim room is counterproductive. Your attention doesn't always stay on the task but wanders into the darker surroundings. Ambient lighting and natural lighting should balance the hardworking area to diminish eye fatigue. A balanced combination of lighting also offers a psychologically gentler environment in which to work and a more aesthetically pleasing room.

The most common hardworking areas are the kitchen, home office or study, and bathroom. A corner of the living room, however, can be turned into a workspace by installing a desk and computer. A garage or basement can be a carpenter's haven. The following discussion addresses the first three generic spaces but offers suggestions that can apply to your home's own hardworking areas.

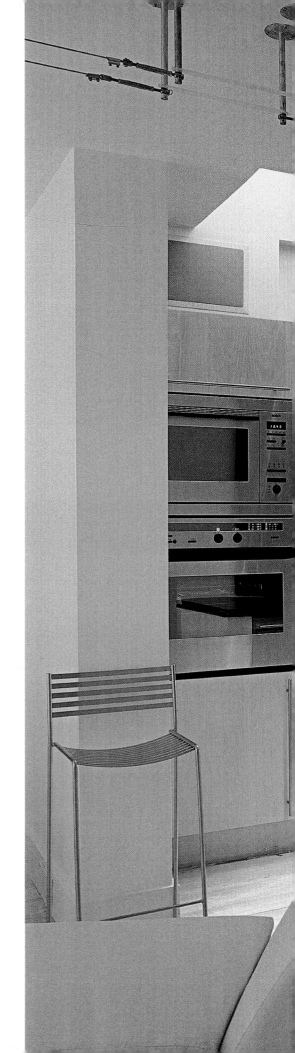

right
Most kitchens function as various zones—areas where track lighting acts as general lighting and others where more concentrated task lighting is necessary. Under cabinet lighting provides an attractive concealed lighting source for hardworking surfaces.

THE KITCHEN The hardest-working area in most homes, the kitchen, is where we store, prepare, and cook food daily. In addition, it's often where the family congregates for casual conversations or for meals. Lighting needs, therefore, include both task and ambient applications. In the food preparation area, the lighting plan is largely fixed due to the permanence of cabinets and appliances. Safety is a major consideration and effective task lighting is essential where sharp implements are used, water is boiled, gas is fired, and electric rings glow. An unnecessary shadow or moment of glare can cause injury. Further, the color the light casts significantly affects our perception of the food we are preparing.

above
A creative use of task lamps in the kitchen.

left
No sleepy souls in this highly reflective stainless steel kitchen. Because of the surface reflectivity, a little light goes a long way and there are obvious and hidden light sources throughout the room.

The limitations in the kitchen also provide excellent solutions. In the prep area, freestanding lights and pendants can be obtrusive but downlights, recessed lights, and lights mounted on the lower surfaces of cabinets are often ideal for task lighting. These lights should be positioned towards the front edges of cabinetry rather than the back, since bulbs placed towards the rear of a cabinet will reflect off the wall and may not reach the part of the counter where tasks are executed. However, lights too far forward can also cause your body or arms to cast shadows on the work at hand, which can compromise safety. So it is important to get the positioning of these lights right, and it often calls for old-fashioned trial and error.

right
Artificial lighting from pendants, under cabinet spots and a fluorescent strip, takes over when the natural lighting dims.

left
Nothing shy about the light and color in this kitchen. The band of fluorescent lights set in the cabinets is supplemented by over head spots for more demure lighting levels for dining.

Concealed Lighting

Task lighting in kitchens demonstrates a decorative trick that can be used elsewhere in the home to dramatic effect—concealing the light source. Take this concept of concealed lighting from the kitchen to other rooms in the house. It's particularly useful for overhead lights and where special shelving display opportunities exist. Use your imagination to identify unlikely places to recess lights. Your task is made easier because of the wide range of bulbs and fixtures that can be used as concealed lighting—incandescent and fluorescent alike.

- Run a strip of fluorescent or halogen lights underneath cabinets or shelves to illuminate a work area.

- Place spotlights underneath cabinets and over work or preparation areas in any room to bring shadowless light to detailed work.

- Use the top of the cabinets or shelves, or even a large piece of furniture such as an armoire, for a run of lights. These reflect off the ceiling to produce a great deal of ambient light.

- Install automatic lights inside cabinets or closets that switch on when the doors are opened.

- Place a light strip underneath a shelf to illuminate the objects below it.

Over cook tops or stoves, lights can be installed under exhaust hoods or on the ceiling, either recessed or mounted on a track and pointed at the work area. Sinks and drain boards also need to be well lit, even if this seems unnecessary because they are ofttimes situated in front of large windows. But it is important to be able to see if dishes are clean, or target the spot where boiling water will be drained, and when it's cloudy there must be enough illumination to compensate for the lack of natural light. Also keep in mind that high gloss surfaces, such as stainless steel or tiles, can cause glare, while dark materials, such as granite or slate, will absorb light so more fixtures or higher wattage bulbs may be necessary.

__**above**
A track configuration—called a color monorail— brings flexibility and visual interest to this kitchen. Additional lighting is provided by cabinet uplights.

__**below**
Taking full advantage of all lighting opportunities— notice the pendants, cabinet up and downlights, ceiling downlights, and even the lights placed underneath the fireplace's hood.

While pendants are not suitable as task lighting, they can work as a source of general illumination when they are astutely chosen and installed, such as a series of fixtures hanging in a row, calculated to be at the right height to avoid getting in the way. Or, they can add a decorative dimension to eating areas, particularly when the dining table is separated from the prep counter. And for ambient lighting, try tube lights concealed behind the moldings at the top of wall cabinets to wash the ceiling with light.

THE HOME OFFICE/STUDY The home computer has become a ubiquitous feature of the American household. Whether setting up a home office or simply turning a corner of the living room into a work area, the computer needs a permanent space. The user needs to comfortably see the keyboard, screen, and notes on the desk alongside. Task lighting—a concentrated pool of light—is essential, but tricky. The task light should be angled to fall on the keyboard, not the screen. At the same time, the light should fall on the sheet of paper without producing glare. The light source itself should not shine in the user's eyes but be protected from view.

Fortunately, task lighting with flexible arms and heads that can be adjusted from a tight, close focus to a more diffuse illumination is available. The most effective task lighting usually takes the form of a flexible desk lamp, which has the advantage of being totally adjustable. It should be able to rotate, bend, swivel or turn to throw light wherever it is most needed. The best known example of this is the standard Anglepoise desk lamp, which is a lamp with a long, bendable jointed arm that comes in myriad variations. Some have heavy bases, while others clip or clamp onto a surface to anchor their arms. Still other directional lamps are flexible thanks to sliding or hinged arms, moving joints, flexible stems, or adjustable shades. With these options, as you move, the light can move with you.

left
A corner of a bedroom can serve as an office, particularly when task lighting designates the workspace, but also when the opportunity for natural light is offered as a relief for the user.

It is also important to correctly position directional task lighting so you avoid working in your own shadow. To do so, place lamps at the opposite side of the work surface from your dominant hand and far enough away do that the light falls across the work surface diagonally from the top left or right (depending on which of your hands is dominant). Also make sure that no light shines directly on a computer screen, which will cause an irritating glare. Direct sunlight can also interfere with the computer screen; the answer is to angle the screen away from the window, and invest in a translucent blind or sheer curtain to filter the sun's rays without blocking them out.

Don't count on task lighting satisfying all your needs, or you will fatigue quickly. Remember that the human eye adjusts more easily from bright light to diffuse ambient light rather than dark surroundings. Therefore, it's best to surround yourself with ambient lighting, particularly lighting that does not cause glare. Uplighting is highly recommended for areas with computers because it incurs no risk of reflection obscuring the visibility of the screen.

_right
Balance task with the general lighting to create a comfortable work area. Place your computer screen slightly askew to windows and artificial light to prevent unwanted glare and make use of window treatments to control natural light levels.

THE BATHROOM Lighting in the bathroom is both practical and
pampering—practical in that significant hygienic activities take
place there every day, and pampering when the lights are turned
low and the bath becomes a place of comfort and refuge. Plus, with
its inevitably shiny surfaces, this room may need well-planned general illumination to soften the hard edges.

The light around the sink and mirror needs to be bright and crisp
for shaving, washing, and applying makeup. It must illuminate the
front and sides of the face, and shine onto the face, not into the mirror.

Thin tubes of fluorescent light, either frosted or hidden behind a
louvered baffle, or a line-up of frosted incandescent bulbs placed
alongside the mirror can be the answer. However, overly bright
lights may be too harsh. Recessed lighting in the ceiling may be
another option and offer a flattering reflection, if it is properly positioned. A light placed directly above a mirrored cabinet can sometimes throw shadows over the surface beneath it. Often lighting in
this area can be more effective when it is concealed behind a protruding mirror and bounced off the ceiling, counters and walls, or
used in tandem with strip lights or bulbs on the sides of the mirror.

right
Sunlight filters in and bounces off the highly reflective surfaces in
this warm-toned bathroom.

Ways to Counteract Glare

Close work requires intense light balanced by ambient light. Glare is an unpleasant and sometimes dangerous side effect of the brightness desired to perform tasks in hardworking areas. Glare is usually caused by bright sunlight streaming through an exposed window or exposed sources of artificial light, incorrectly designed lighting fixtures, too much light from one direction, or excessive contrast between lighted and shadowed areas. To control glare, follow these tips:

- Curtain windows with sheer fabrics or blinds that either pull up or down to half-shield the glare of the natural light.

- Correctly shade bright sources of artificial light to diffuse or direct its brilliance. When using a directional lamp, always make sure the bare bulb is completely concealed since even a bit of exposure will produce an irritating glare.

- Bring light from two or more directions.

- Provide a low level of general illumination to diminish contrast in a room with strong task lighting. For close work, the working area should be no more than five times as bright as the darkest part of the room. A ratio of one to ten is never desirable.

Aside from the sink area, the bathroom lighting need not be overtly strong—just bright enough to accomplish daily tasks. Recessed downlights can create subtle pools of light that reflect off the tiled shower walls or mirrored surfaces. A dimmer switch can set the mood for that soothing bath, or allow the lights to be turned up for reading in the tub. Skylights can bring in natural light. A small light inside the cabinet, triggered when the door is opened, can be convenient for reading the labels of cosmetics and medicines. Whether a toilet is sequestered in its own cubicle or open to the rest of the room, it should also be well-lit.

Safety is an issue in the bathroom. Many fixtures are specifically designed for use in proximity to water. In general, freestanding, hanging, or adjustable light fixtures should be avoided, as should standard sockets, switches, and cords. A night light strong enough to make it possible to fill a glass of water or use the toilet should also be present for nocturnal visits to the bathroom.

above
The understated design of this bathroom is matched by its simple lighting solution.

left
Enjoy the old-fashioned bathing experience complete with an abundance of warm daylight filtering through blinded windows.

> **Types** of Light

The term lamp is usually used to refer to freestanding fixtures but, technically, a lamp is what is most commonly called a light bulb. The bulb itself is merely the outer envelope of an artificial light source, which contains a variety of gases and other elements that produce correspondingly varied light. Those components, and the way an electric current is passed through the bulb, gives the light its quality and color and qualifies it as either fluorescent or incandescent.

> Incandescent Lamps

These are simple devices that consist of a wire, called a filament, sealed in a bulb that is filled with gas. An electric current passing through the wire heats it until it incandesces, or glows. The diameter and length of the filament determine the amount of electrical current, or wattage, that is consumed by the lamp, and regulates its light output.

There are many types of incandescent lamps, all of which use tungsten wires sealed in bulbs filled with several gases. Incandescent lamps produce a yellow light that is most like the sun and accents the warm colors of the spectrum. However, the colors in a room can also look a little yellower than they really are under this kind of lamp. Most are inexpensive, but also don't last very long and start losing brightness as they age. Two types of incandescent lamps produce whiter lights and have longer lives: full-spectrum incandescents and halogen lamps.

Full-spectrum incandescent lamps emit a slightly whiter light that is closer to daylight and gives a truer, more vibrant cast to colors, which can be quite uplifting in a dreary space. But they can also make a room inside a house feel like it is outdoors. They are more expensive than regular incandescent bulbs, but last longer.

Halogen bulbs are filled with the gas halogen instead of the more common argon, nitrogen, or krypton. They produce a whiter brighter light that accents colors more accurately unless dimmed, when the light becomes warm and reddish. Halogen bulbs are smaller and more energy-efficient than other incandescent lamps, and last about three times longer. But halogen lamps have many drawbacks. They produce a small, intense light from a tiny bulb, so shadows are sharper and harsher reflections make it more difficult to see. They can also get very hot, sometimes radiating so much heat that particles of dust or dirt on them smoke; they can even cause fires. Finally, these lamps are extremely fragile, much more expensive than regular incandescent bulbs, and may be difficult to replace, depending on their size and shape.

> Fluorescent Lamps

A fluorescent lamp produces light by passing an electric current through a vapor or gas rather than through a tungsten wire. This is called a discharge source and is a much more efficient method than that used to heat filaments in incandescent lamps. Though there are several types of discharge lamps besides fluorescents, such as high-intensity discharge (HID) lamps, these are not suitable for residential or interior uses because they take quite a bit of time to warm up and emit light.

Fluorescent bulbs are much more expensive to purchase and install than incandescent bulbs, but they are also much more economical in the long run because they produce at least three times the light for an equivalent wattage, last ten times as long, and emit less heat into a room, which is advantageous in warm climates and the summer months. Their light distribution is also more even, and the intensity does not diminish with age. Fluorescent bulbs produce a cool white light whose quality is similar to natural light and is ideal for ambient illumination; this is fine by day but can be a drawback at night, as it can make a room feel bright and unnatural.

However, warmer full-spectrum fluorescents called compact fluorescent bulbs are on the mar-

—**above**

Nothing shy about the light and color in this kitchen. The band of fluorescent lights set in the cabinets is supplemented by over head spots for more demure lighting levels for dining.

ket. Though they've been available for about a decade, recent improvements in the design and color resolution of these bulbs have increased their desirability for home use. Now their color-rendering properties are similar to the warm, mellow tones of incandescent light, and they come in shapes and sizes that are effective for home use. Though initially more expensive, these fluorescents save money in the long run because they last six to eight years (which makes them ideal for hard-to-reach places), use one-quarter of the energy of incandescent bulbs, and don't pose the same hazards as halogen lamps (which are a fire hazard).

Best of all, the savings are substantial when using this type of bulb. A 23-watt fluorescent bulb replaces a 75- to 90-watt incandescent and costs about $10 to $12, but it burns for 6,000 to 10,000 hours (or three years, based on eight hours a day of usage) and saves $20 in energy costs a year. Multiply that $20 by fourteen, which is two bulbs per room in a seven-room home, and you've saved $280 a year. Spread over three years, that's over $800 for an initial $8 to $12 investment.

> **Window** Coverings: Controlling Natural Light

We strive to bring natural light into our homes, but sometimes the light is too bright or the reflection too distracting. Control of natural light is particularly important in work areas, where interior light levels are highly concentrated.

Window treatments are the answer. These solve many other problems as well; they limit both summer heat gain and winter heat loss, screen for privacy or eliminate an unpleasant view, improve the window's appearance, and introduce color or texture into the room. Window treatments are also flexible in that they adjust to the changing natural light conditions of day and night. Here's a basic list of window treatment types:

> **Blinds or shades:** These come in several types. The generic blind or shade is a piece of cloth or heavy paper on a spring roll; it is usually pulled down from the top but can be made to open from the bottom or side.

> **Roman and Austrian blinds:** Considered traditional in design, these matchstick or slat blinds are usually thin bamboo or wood strips that pull up into a roll.

> **Venetian blinds:** These come in horizontal (most common) and vertical designs and are made of wood, plastic, or aluminum in a variety of colors and finishes. Contemporary design favors blinds with narrow slats (5/8 inches).

> **Thermal shades:** These energy-saving shades are made of plastic sheet or special weaves that provide sun controls through tiny microlouvers. Light and heat sensors can adjust thermal shades for optimal benefit.

> **Drapery:** This broad category includes any loosely hung fabric that covers an entire window and extends from floor to ceiling or wall to wall. Sheer fabrics permit some light transmission. Heavier fabrics or several layers further control light levels. Drapery introduces aesthetic considerations at window openings, offering color, pattern, and various kinds of textiles.

> **Curtains:** Curtains are a modest form of drapery in that they are usually placed within the window frame and offer limited light and vision control. Styles include sash, café, lace, and net curtains.

> **Shutters:** These moveable structures can be solid or louvered and may be used in combination with shades, curtains, and drapery.

> **Shoji screens:**
These light-transmitting screen panels slide on tracks and are based on Japanese tradition.

above
Pendants hang over the breakfast bar and cooking island and contribute to the sleek style of this breakfast nook. A narrow-slatted, pull-up blind controls the amount of natural light that enters the room.

Hanging lanterns always seem to evoke a feeling of a simple, more romantic time, especially when they're hanging from a beautiful hook. You can make one or more of these in a palette to suit any mood or room, with readily available and inexpensive materials. Any small glass with a lip for the chain to be fastened to will work well. Try using different styles of chains, painting several differently shaped glasses the same way, or painting the glasses a solid color.

Marbled Hanging Lantern

Materials
- Gallery glass paints in the following colors: clear frost, denim blue, royal blue, and turquoise
- Glass container with lip
- Jewelry chain
- Five silver-tone jumprings
- A closed loop for hanging, such as an earring jewelry finding
- Water-based acrylic craft sealer
- A soft paint brush
- A wire cutter or old scissors
- String
- Two pairs of pliers

TIP To ensure that you don't accidentally touch the areas you've already covered, try painting the lantern upside-down on a lazy susan, or set it on a piece of cardboard, which you can rotate as you paint.

1. Fit the glass with the hanging chain.
Wrap a piece of string under the lip of the glass to measure the circumference. Use this as a guide to measure and cut the chain. Then, cut the chain in half, and connect two ends with a jumpring. You can do this by using two pairs of pliers to pull apart a jumpring where it is separated, then pulling it closed once the chains are hooked through it. Secure the chain to the glass with a piece of masking tape to hold it in position, loop another jumpring to the other end of the chain, and test the fit. Then remove the chain.

2. Paint the glass container.
This paint sets quickly and becomes translucent after it dries completely. I recommend getting used to the paint and the painting technique described below first by painting a bottle or jar.

Squeeze out dime-sized amounts of paint on the surface of the glass, beginning with the clear frost, then use a toothpick to swirl the colors together until you have the desired marbled effect. Work in small sections, blending the edges of the area you've just painted with the new paint you add, and be sure to spread the paint out evenly in a thin layer. Paint back and forth, rather than circling in one direction, so that the area you painted first doesn't dry before you reach the edge of it. Once the paint begins to set, it can become lumpy if you try to marble it. The paint can be washed off easily before it dries, if you want to start over. Let the paint dry overnight, then apply two coats of sealer to protect the finish.

3. Attach the hanging chain.
Cut two equal lengths of chain. Open the jumpring that connects the collar that you've already assembled for the glass, loop it around one of the chains you've just cut, and close it off. Then, position the chain around the lip of the glass and use the tape to hold it in position. Using another jumpring, connect the ends of the lantern's collar, then loop the final chain in the jumpring and close.

Add a jumpring to the two loose ends of the chains. Use the last jumpring to connect them to the hook or loop for hanging.

Henna patterns, used to decorate the hands and feet of Indian brides, often have an appealing, highly stylized natural theme such as the flowers on this lampshade. With a permanent fabric marker, you can easily reproduce the color and delicate linework of this traditional art form. A thin but tightly woven fabric takes the marker best. You can create your own henna patterns by consulting one of the readily available sourcebooks or kits for doing your own henna body art.

Henna Pattern Lamp

Materials

- 1/4 inch (.5 cm) yard of off-white cotton fabric
- 1 package of 1/4 inch (.5 cm) double-fold bias binding, off-white
- Self-adhesive lampshade, 3 inch x 5 inch x 4 inch (8 cm x 13 cm x 10 cm)
- Marvy Uchida fabric brush marker in brown
- Fabric transfer paper
- Fabric glue
- Scissors
- Simple black lamp base that takes the smaller "candle" bulbs

TIP To easily experiment with different designs or to see how something will look, cut out a piece of paper in the shape of your shade, sketch the design on it, and tape it over the shade's protective wrapper. You can also make your own complete pattern this way.

1. Cut the fabric for the shade.

Prewash and iron all fabric. Tape the fabric to your work surface, and trace the shape of the lampshade wrapper on the fabric with a pencil. Cut the fabric along this line.

2. Draw the pattern onto the fabric.

Before beginning to draw the pattern, get used to the fabric marker on a scrap. The thickness and thinness of the lines are controlled by the amount of pressure you use, and you can produce calligraphy-like lines.

Copy the pattern to be 2 1/2 inches (6 cm) tall. Secure the paper and fabric with tape. Transfer the pattern to the fabric using fabric transfer paper, or trace over a light box; space four of the images about 1 inch (3 cm) apart from each other. Do not draw the border yet. Use the brown fabric brush marker to draw the design.

3. Attach the fabric to the shade.

Starting at the seam, wrap the shade with the fabric. Fold the other edge under, securing with a little fabric glue, then use the glue to secure the seam.

4. Add the trim.

Cut 9 inches (23 cm) of trim for the top of the shade and 18 inches (46 cm) for the bottom. Using fabric glue, secure it to the edges of the shade, inside and out.

5. Draw the border at the top and bottom.

The fabric stretches once it's attached to the shade, so it is difficult to get a border to match up. The easiest way to do this is to freehand draw it on after the fabric is attached.

Draw a line around the top and bottom of the shade about 1/8 inch (.25 cm) from the trim. Then, draw dots close to the lines all the way around, then follow with another line after the dots. Finish with little "commas" with the tails ending at the lines.

Lighting All Around the House

From the entrance to the house to the most hidden corner of the basement closet, consistent light levels are an essential ingredient—and often taken for granted. Here is one scenario: A set of decorative lanterns guides you inside your house. You flick on the recessed hall lights without concern for what might be waiting in the dark. You pass through the brightly lit hallway to the basement stairs.

As your foot hits the first step, the movement sensor trips the light switch, turning on the fluorescent ceiling light below. You hurry to the far corner of the basement without a thought about how easily you are finding your way in the once-dark cellar. You open the closet door. A small halogen bulb switches on. Deep in the closet are the tennis balls you almost forgot to take. You turn around and, as you go, the lights switch off behind you. The progression is automatic.

We need light in hallways, foyers, stairs, and basements to safely negotiate our way. The trick is to reach beyond the purely practical to the aesthetic. Meet the need for light throughout the house while making a theatrical statement or generating a warm and relaxing atmosphere. Rather than just lighting a staircase from above, position a sequence of recessed wall lights so that the light grazes each stair tread and riser to create a playful visual image. Use lighting to create a sense of progression at the entrance or to heighten the exterior architectural drama. Think of a long, narrow passageway as an opportunity for display with downlights or uplights. Use spotlights to turn a garden into a nighttime room. Light can be used to enliven the dullest spaces.

left
Add surprise and delight to a room with decorative lighting.

LIGHTING SPECIAL-NEEDS AREAS (HALLWAYS, FOYERS, STAIRS, BASEMENTS) When we think about lighting our home, we overlook some spaces that get the most use—the hallways, foyers, and stairs. And not only do we forget how critical adequate illumination can be in these spaces, we tend to think of them as areas where function takes precedence over form. In fact, both count in the case of the hallway and foyer, and it is so easy to use lighting as a decorative tool, why not try to make the basement as attractive and charming as possible?

The foyer is the space guests see first. A half- or indistinctly lit foyer makes it harder to engage the warm note of hospitality than one gently illuminated with ambient ceiling or wall lights. If the space is used to display photographs, artwork, or collectibles, consider spotlights or track lighting. Use recessed downlights or directed spots to accentuate the path from the front door to the rest of the home. Create a sense of drama and architectural distinction with wall-mounted uplights. Any of these techniques, alone or in combination, allow you to welcome your guests and make it easy for them to admire your home.

___**left**

When natural lighting isn't enough, supplement your workspace with well-placed accent lamps.

The hallway need not have the most pedestrian style in the home. An antique ceiling light can add a decorative flourish, as can pendant lights, if the ceiling is high and the lights are not hung too low. Sconces that protrude from the wall on arms or branches reflect on painted woodwork and polished floors. Recessed downlights mark the path, as do wall-mounted sconces or uplights. If the hall is wide enough, a lamp on a small shelf or table in a corner is a pleasant touch. If possible, add skylights to the hall's ceiling to bring in natural light during the day, or lighten the hallway with a strategically placed mirror that will expand the space and reflect illumination.

The most common way to light the stairway is with ceiling lights at the top and bottom. Avoid that old-fashioned approach by installing wall-mounted or recessed lights along the treads at foot or shoulder level. Lights at foot level work best for safety reasons; shoulder-level lights are for aesthetic effect. Be careful with angle or pendant lighting, which can cause glare. A skylight at the top of the staircase can add a dramatic natural effect.

above
During the day, natural light filters into the hallway through the glass block wall inserts. At night, the hallway is illuminated by uplights placed in the floor.

right
Bookshelves offer an unexpected lighting source. Place tube lights underneath each shelf to brighten up a cove or a dark hallway.

Basements often serve many purposes—storage area, laundry, playroom, workshop, music room. The new shapes, sizes, and color renditions of fluorescent lights find utilitarian use in basements, where the value of long-lasting light outweighs the desire for pure aesthetics. Ceiling or wall-mounted lights work best in places of high activity. Incandescent lighting should also be considered in the basement, since it can bring softness to a hardworking space.

above
Increase safety and beauty by lighting stair treads. Simply place side-lights in the walls next to the stairs.

left
Brighten stairs by inserting sidelights at tread level.

LIGHTING OUTDOOR AREAS Outdoor areas are lit for security, but also for safety and beauty. The larger the house and the more isolated the location, the more lights are generally required. In urban areas, street lighting can provide some visual aid, but probably not enough to allow you to get the key in the lock on a dark night or appropriately welcome visitors. However, overlighting can also be a problem, since it can cause illumination or glare to spill over into the windows of nearby houses.

Effective exterior lighting doesn't necessarily have to be bright. Spotlights can be used to illuminate the pathway to the house, along either the driveway or the front walkway. Carriage lamps or lanterns at the front door warmly beckon to the visitor. Position the lights carefully to avoid glare. A grand building façade may be lit with several spotlights focused on the most impressive architectural details or contrasts in textures or patterns. Flooding the façade dulls the building's appearance. Locate lights at low positions and create pools of light.

right
Selectively placed hanging lanterns and table lamps create a room outdoors and extend your living space.

Gardens offer a delightful stage for lighting. Back light foliage, silhouette garden structures, spot light ornamental pots, all for theatrical play and variety of scale and proportion. Ornamental lighting can be provided through uplights inserted in the ground or lanterns set on the ground, or even downlights. Use higher lighting levels in the barbecue area—or at least provide for stronger lighting when food is being prepared.

For safety reasons, consult a qualified professional when planning outdoor lighting installations. Outdoor lighting needs its own infrastructure, and should be installed by a professional. Fixtures and wires are constantly exposed to weather, extremes of temperature, and lawnmowers and garden tools, and must be safe and sturdy. Switches should be operable from indoors.

left
Exterior lighting should spot distinctive features of your house, as well as provide paths of access and a warm welcome.

> **Turning** a Garden into an **Outdoor** Room

Indoor-outdoor living is popular, especially among people who live in a small house or a mild climate. Lit creatively, a garden patio can be a glorious extension of the home, offering the grandest nighttime entertainment spot. A lighting scheme that creates a certain mood makes or breaks the space. Bright floodlighting won't produce the intimacy desired. Rather, think of the pools of light that you try to create indoors. Here are some suggestions:

> **Concentrate** on soft, glowing light levels.

> **Start** with candles, torches, and lanterns.

> **Use** uplights and accents around the patio directed toward the surrounding plantings to dramatize the different shapes and textures of the plants.

> **Use** small spotlights, like string lights used for Christmas trees, to highlight an area where food is served or embellish particularly attractive trees.

> **Section** off parts of the yard with borders made of votives, but position them so they won't be easy to knock over.

> **Create** more than one focal point for your outdoor room to express the pools-of-light idea.

> **Consider** how to build the light from dusk-to-dark night in your outdoor room to create a sense of heightening drama. You can do this by turning the lights on in various areas in stages.

> **Build** flexibility into your outdoor room so that it can be lit differently from time to time.

> **Create** a structural presence in your room with a trellis or arch, which can be imaginatively lit from below.

> **Line** pathways with votives on the ground, placed out of the way of feet, or string lights draped along low bushes or trees.

Safety First

Safety is a double-edged sword when it comes to lighting. We need light to make our way safely around the house, yet at the same time, we need to be aware of the safety issues surrounding light and electricity. Here many ways to cope with potential hazards:

- Light stairs, halls, and passageways well, so you can get safely around your house.

- Avoid overly bright lighting that can cause glare and deep shadows. Light all stairwells. Illuminate exterior entrances adequately for security purposes. Use timers to operate lights when away from home.

- Don't guess at electrical installations; consult a professional for indoor and outdoor wiring.

- Make sure you provide enough ventilation room with recessed lighting. The hotter the bulb, the more ventilation it needs.

- Make sure water and electricity do not come in contact in bathrooms and kitchens.

- Watch for faulty wiring or plugs and overloaded sockets.

- Light bulbs get hot, especially halogen lamps. For instance, the heat from a 300-watt halogen bulb can top 900 °F and scorch surfaces 18 inches away, so don't place these lamps near surfaces they can damage.

- Don't place lamps near drapes, curtains, or wall hangings that could billow in a breeze and brush against them or tip the lamps over and be ignited from the heat.

- Don't' leave hot bulbs on for long periods in room unattended, as they can be a fire hazard if they shatter, tip over, or scorch something nearby that ignites.

- Don't leave anything draped over a lamp arm or lampshade (either accidentally or purposely, such as a piece of cloth to change the lamp's mood), as it can catch fire from the heat of the bulb.

- Don't look directly at a powerful lamp. The intense glare and heat, especially from halogen lamps, can harm your eyes.

> **All** About Bulbs

Bulbs are often referred to as lamps, which is the technically correct term for the glass that encases a filament or holds some type of gas and glows when electricity is applied.

Information on the shape and size of each lamp or bulb is imprinted on it. Shape is identified by a letter, while size relates to the bulb's diameter at its widest point and is identified in 1/8-inch multiples. For instance, an A-21 bulb has an A (standard) shape and a 21-inch by 1/8-inch (or 2 5/8-inch) diameter.

Make sure you know the shape and size of the bulb that is needed before going to the store. When shopping for a new fixture, it pays to take bulb shape and size into consideration, as some bulbs are hard to find or must be custom ordered. This is particularly important with imported fixtures. Finally, when buying a fixture for a hard-to-reach spot, consider one that can use the new long-life compact fluorescent bulbs, which need not be replaced very often. When possible, install dimmers, as

reducing the level of light by as little as 10 percent can increase the life of a bulb and sometimes even double it and provide more lighting level options.

The most common type of incandescent lamps is the standard frosted or opaque GLS (general lighting service) bulb. Silvered-bowl GLS bulbs, the simplest type of incandescent reflectors, throw light back into specially designed fixtures or are used to prevent glare in pendant fixtures. Opal white globe bulbs are also quite common and emit a soft, diffused light that doesn't cast hard shadows. Candle-shaped bulbs, which emulate the glow of candlelight, come in several slender shapes that are frosted or clear, depending on whether or not they are to be used with shades (the frosted variety is intended to be used without

shades). Parabolic aluminized reflector (PAR) economy bulbs have a sturdy front made of thick, textured glass that enables them to withstand higher temperatures and are available as incandescent or halogen lamps. Reflectors come in two versions: a spot that emits a beam of light less than 30-degrees wide, and a flood that emits a beam of light greater than 30-degrees wide.

Halogen lamps burn at much higher temperatures than regular incandescent lamps, so the bulb is smaller and made of quartz to withstand the higher heat. Standard household-current halogen bulbs are used in conventional fixtures; tiny, compact low-voltage halogen bulbs require transformers to reduce the power. Low-voltage halogen diachronic reflector bulbs, which must also be used with transformers, direct light forward and draw heat back, which creates a concentrated beam of cool light.

Fluorescent lamps have been traditionally used in commercial applications, thanks to the cool white nature of the light they emit, but they are now available in compact sizes that have warm color-rendering properties closer to that of incandescents. They cost more in the short run but last six to eight years, use approximately one-quarter the energy of regular incandescent bulbs, and come in a wide range of shapes. Now there are fluorescent bulbs that emulate the standard form of the familiar GLS incandescent bulb and tread new ground with their innovative designs that sport coiled, slim, and circular shapes. Even so, they fit a variety of standard fixtures and provide cheaper, brighter, and more efficient light. While there are no hard-and-fast rules when it comes to choosing a type of bulb, some helpful guidelines are available:

> **As a general rule,** the lamp in a fitting should be shielded from view because it can be extremely irritating, distracting, and harsh on the eye. Overly bright, badly positioned bulbs can actually cause temporary blinding.

> **Certain light bulbs** have a specific decorative purpose. Flame-tipped bulbs, which are also called candelabra bulbs, simulate candles, and, if frosted, can remain exposed in sconces. Tinted bulbs can be used to give a warm glow to a room, cool down its colors, or dress it up for a special occasion. While incandescent bulbs come in a variety of tints, fluorescents can be covered with colored sleeves or manipulated with sheets of cellophane gel (which are clipped over the opening in a shade to tint the emitted light). Both of these products are routinely sold at lighting or photographic supply stores.

> **Reflector bulbs come** in floodlights and spotlights, and have a reflective silver coating inside the glass to direct light forward and provide better beam control than general service bulbs. When compared with a general purpose bulb of the same wattage, this treatment can double the amount of light (in foot-candles) thrown on a subject or spot.

> **Handling a bulb** shortens its life because dirt, grease, or moisture from fingers can cause it to shatter once it is turned on (halogen bulbs are particularly susceptible to this). Use a cloth or paper towel when installing bulbs, first making sure the fixture is turned off at the power source

High-Tech Lighting Accessories

We're all familiar with the automatic switch that turns on the refrigerator light when the door is open and off when the door is closed. These switches also work well in hall closets and other storage areas. To give the impression of occupancy, install timers and time switches on lights in the kitchen and other heavily used rooms. More advanced are approach lights that can be triggered by heat, motion, or sound. Some switches are tripped by infrared beams, which are good security devices. It is likely that the fully computerized home of the future will be equipped with photosensitive cells that switch lights on and off in response to natural light levels or the approach of a potential user. Or the lighting of the whole home may be programmed to turn on or off at certain times. The goal is a more efficient and economical use of energy. Some of this technology is already in use in theaters and large commercial and public spaces.

>**All** About Lamps and Shades

Like the right piece of jewelry or scarf, a lamp can make a major impact on the style quotient of a room. Plus, an astonishing range of interesting and innovative lamps is on the market today, ranging from intricate antiques or reproduction versions to cleverly conceived contemporary designs that employ materials in new ways and boast sleek or sculptural lines. In fact, some of these are functional works of art in their own right (see Decorative Lamps Tip.)

Lamps can also be made easily from a wide range of objects, such as vases, jars, and coffee- or tea-pots, thanks to simple, do-it-yourself adapter kits readily available at craft, hardware, and lighting stores. Almost any object can work as a base, as long as it is stable, hollow, and has center holes through both ends (but don't drill holes in an antique, as this destroys its value).

The shade is an often overlooked but critical component of a lamp. Shades that are dark in hue glow softly when a light is turned on, those made of translucent materials cast more light into a room and make a space seem brighter, and opaque shades direct the light rather than emit it. But all shades dictate where light falls by their shapes. Those with a wider diameter cast a larger pool of light than narrower shades, though repositioning a shade with respect to the bulb changes this equation, as the closer the bulb to the bottom edge of the shade, the larger the pool of light.

Besides regulating the way light is dispersed, shades can make or break the lamp aesthetically. Take the lamp base to the store and try it out with various shades before making any decisions—especially if the lamp is an important accessory in a room. Though one formu-

la dictates that the diameter of the bottom of the shade should be equal to the height of the base, this approach is unreliable. The ultimate success of the marriage of base and shade depends on the shape, style, color, and composition of both parts of the lamp. When choosing shades, keep the following points in mind:

> **Consider function first.** If a lamp is intended for task lighting, use an opaque shade that directs light down; if it is intended for general lighting, use a translucent shade that casts a broad and soft glow. Shape also affects how light is dispersed. The three main shade shapes are drum, which is cylindrical; empire, which is slightly wider at the bottom; and coolie, which resembles the hats worn by Chinese peasants and has broadly sloping sides.

> **Pick a style that suits the decor of the room.** Checkerboards and stenciled motifs evoke a rustic or country demeanor, muted florals or simple laces can be unabashedly romantic, pleated or plain silks can be elegant and formal, and textured or grainy papers can be contemporary and sophisticated.

__**above**

Choose a lamp that expresses the decorative mood of a room or a painting near by.

> Pick the right size.
Relate the shade to the scale of the base first, the scale of any surrounding furnishings next, and the scale of the room last. The visual weight of the lamp base, not its actual dimensions, should be about two-thirds the entire lamp, with the shade making up the remaining third. Between the shade's bottom edge and the surface it rests on, the distance should be greater than the shade height so the lamp doesn't look top-heavy. The shade should also extend at least two inches beyond the base on all sides.

> Check the positioning of the shade on the lamp base.
Shades feature one of two types of carriers: a metal structure that attaches to the base, or a metal structure that hugs the bulb. In both cases, the shade should hide all of the metal working parts of the base when viewed from eye level.

> View the shade with the bulb illuminated.
The color, material, and shape of the shade have a profound effect on the quality of light the lamp emits, so make sure the shade fulfills your wants and needs before purchasing it.

right
Whenever possible, let natural lighting be the main source of light. Then, just supplement your space with fun lighting, as with the demi-wall with twinkle lights.

Decorative Lamps:
Style over Substance

Consumers are asking a lot of lighting designers these days. Not only do we want functional lights, we desire aesthetically pleasing lamps. And manufacturers are responding with everything from high-tech floor lamps to antique replica chandeliers. Purely decorative lamps, however, go a step further. These lamps display light as an art object and are quirky, fun, playful. Often the lamps are not bright enough to read or do hard tasks by, but they are wonderful to look at and magical to have in one's home.

■ Perhaps the easiest way to envision a decorative lamp is in the children's world. There are carousel lamps that project cartoon images of favorite characters, animals, stars, or plants onto the walls; ceramic figures or colorful paper shapes that light up; and lamp bases made of actual toys or pieces of sports equipment, ranging from baseball bats to dolls.

■ Decorative lamps for adults are much more abstract, relying on the pure imagination of the artist. The light can move, like the flickering of a candle, or shine intensely, like neon gas. It can sparkle behind a sculpture representing a leaf or an abstract globe. Tiny lamps can produce jewel-like sparkles.

■ Decorative lamps are sometimes available at lamp stores, mixed in with more conventional lighting. They are more likely to be found at arts and crafts shows, art galleries, and artist studios. Decorative lamps can add an unusual element of delight and artistry to your home.

left

Two contemporary versions of the decorative lamp.

above

Consider light as sculpture. Light can have an artistic purpose as well as a utilitarian one, as seen in this playful bedroom wall hanging, which supports the popular "points of light" metaphor.

A simple shell can shed a warm, soothing light and soften the look of a utilitarian night light, and the easily added pearlescent foiling adds a natural-looking shimmery effect. When selecting a shell, test it against a bulb to be sure it is translucent enough. Rather then using a single shell, you could try gluing small, thin shells together in a fan shape.

Pearled Shell Night Light

Materials

- Standard night-light, with removable front
- Shell
- Delta Renaissance Foil in mother-of-pearl
- Delta Renaissance Foil adhesive
- Delta Renaissance Foil sealer
- Soft brush
- Hot glue gun or strong general-purpose adhesive

TIP Beautiful shells that are large enough to serve as a night light shade may be difficult to find. I used a simple scallop shell, which diffuses light beautifully, and are often sold at kitchen supply stores as seafood serving dishes.

1. Apply the adhesive to the shell.

Select a shell that is fairly flat and has an area at the bottom that can be fitted to the night light.

Following the manufacturer's instructions, apply a thick even coat of adhesive to the shell using a soft brush. Try not to go over the same area twice, as this may result in the adhesive becoming textured or lumpy. When the adhesive has dried and become translucent, apply a second coat and let it dry until it has become translucent.

2. Apply foil.

Cut a piece of foil big enough to cover the shell, plus a little extra to leave unattached for easier removal. Lay the foil on the shell and smooth it out gently with your fingers. Using sharp objects will damage the foil, so use something soft for crevices, such as cotton swab or eraser. The shell I used had many grooves and a lot of texture, so I had to go over the shell several times, concentrating on small areas each time. Try not to touch the bare adhesive with your fingers, and be sure not to let the exposed foil backing touch the adhesive, as it will pull it off.

3. Seal the shell.

The foiled shell is prone to scratches and fingerprints, so carefully apply one or two coats of sealer to protect it before attaching it to the night light.

4. Attach the shell to night light.

Fit the shell to the night light to see where the two meet, and put two or three small dabs of hot glue on the shell at this point. Press it to the night light, trying not to let the glue be visible from the front. Add a few more dabs of glue inside, if necessary, and hold the shell in position until the glue firms up, about one minute. If you need to, remove the shell before the hot glue has dried and try again; it should be easily removable at this point.

Hot glue is ideal because of the quick bond and easy removal, but you can use a strong, industrial adhesive. Be sure it is formulated for a variety of surfaces, including plastic.

Two or three of these tea lights can be used for an intimate table setting, or en masse for a festive effect. With a simple foiling system, you can make these in copper, gold, or silver. Try "painting" a simple pattern with the adhesive, such as stripes, rather than foiling the whole votive, or unifying differently shaped votives with the same pattern.

Silvered Tea Light Votives

Materials
- Small glass votives
- Delta Renaissance Foil in silver
- Delta Renaissance Foil adhesive
- Delta Renaissance Foil sealer
- Soft brush

TIP To get the best effect from the semi-opaque foiling, use smaller glass containers such as those used for tea lights, so the flame will be closer to the glass.

1. Apply the adhesive.

Clean and dry the votives. Following the manufacturer's instructions, apply a thick even coat of adhesive to the votives using a soft brush. Try not to go over the same area twice, as this may result in the adhesive becoming textured or lumpy. When the adhesive has dried and become translucent, apply a second coat and let it dry until it has become translucent.

2. Apply the foil.

Cut a piece of foil about 4 inches (10 cm) square, or large enough to cover one side of the votive. Lay the foil on the votive, being sure to leave a small section unattached for easier removal, and smooth it out gently with your fingers. Using sharp objects will damage the foil, so use something soft for crevices, such as cotton swab or eraser. Repeat until the votive is covered. Go over bare spots with more foil, but be sure not to let the exposed foil backing touch the adhesive, as it will pull it off.

3. Seal the votives.

The foiled votives are prone to scratches and fingerprints, so carefully apply one or two coats of sealer to protect them.

Addressing Special Lighting Needs/
Lighting Solutions

The subject of lighting can be needlessly complex. While focusing on technical advice about fixtures and bulbs, we forget to rely on our most reliable tool for assessing lighting effectiveness—our own eyes. It is easy to distinguish between good and bad lighting, or adequate and inadequate lighting, merely by looking at it. When you live in a room and perform a range of activities or tasks there daily, its faults become all too apparent.

Good lighting meets your needs, puts you at ease, creates a sense of comfort, blends into the background, and enhances the colors, forms, and textures of the furnishings in a room. Bad lighting is jarring and leaves you with the impression that something is amiss. It may imbue a space with an uneasy atmosphere or make its occupants jumpy, nervous, and tense. More significantly, it can cause headaches, eyestrain, nervousness, or even accidents due to poor visibility.

left

In an exceedingly narrow room with high ceilings, the headboard was placed on the side of the bed and flanked by wall mounted reading lamps that swivel to different positions.

In the end, the specific purpose and design of a room as well as the wants, needs, and personalities of its users determine the look, intensity, and distribution of its lighting. Fortunately, there are plenty of ways to improve the caliber, quality, and ambience of the lighting in a room without incurring the mess and upheaval of running wires into ceilings and walls. It is possible to change the mood of a room, up its style or drama quotient, or improve its performance with a few simple changes or tricks. Here are some of the simple principles the pros apply in using lighting as a tool throughout the home.

above
Dress up the traditional light shade.

right, above
Eclectic lamps complement the room's creative décor.

right, below
A whimsical interpretation of bedside reading lamps.

LIGHT AND COLOR Artificial lighting can do unexpected things to the colors of furnishings and textiles. Incandescent lamps, which emit most of their light at the yellow to red end of the spectrum, add yellow to everything they illuminate. Accordingly, blue assumes a greenish cast, red an orange cast, and white a creamy cast. Halogen bulbs, which are also incandescent, produce a whiter light, but when set on a dimmer at low, they assume a reddish glow. Fluorescent light is a cool, bright white light that can be annoyingly sterile and white, but the new compact fluorescent bulbs cast a warmer glow that is quite close to the tones of incandescent lamps. To avoid a room that works only at day or night, check the color of upholstery, rugs, drapes, and paint under all the lighting conditions of a twenty-four-hour period. Paint a piece of wallboard or plywood (several feet square) the same color you plan on painting the walls; check it at all times of the day and night to see how the light affects it.

left
A discrete multiple-spot provides diffuse ambient lighting in this sitting room that by day is flooded with natural light.

INCREASING LIGHT BY DAY Dark walls and surfaces, especially those with matte finishes, absorb light. This applies to everything in a room, including walls, window treatments, upholstered pieces, and rugs. At its most basic, increasing the light in a room involves using light colors on surfaces and furnishings. On walls, use gloss or semi-gloss paint; on furnishings, use textiles that have reflective properties, such as satin-finished fabrics and twills rather than nubby or textured textiles. To increase the light that filters into a room, use translucent rather than opaque window treatments. If a space does not get enough light from its windows, steal light from a brighter adjacent room by installing transom or fanlight windows over doors or turning complete walls into three-quarter-height partitions. It is also possible to add skylights or clerestory windows to a room. For rooms facing north, artificial light is necessary during the day. To make its presence less obvious, stick to sources that are close to natural daylight, which is whiter than incandescent light. Full-spectrum bulbs are one option; fluorescent striplighting is another. The latter can work well by day, concealed behind cornices, in the top of a bookcase, or recessed inside soffits above windows.

right
The translucent shoji screen wall allows light to enter from the adjoining room, which can be supplemented by task lighting or candlelight.

INCREASING LIGHT AT NIGHT When determining how much light to use at night, the first consideration is whether the illumination should have a strong physical presence or be more of a background aura. For lots of bright light, use downlights in tracks, recessed cans, or pendant fixtures, supplemented with floor and table lamps throughout the room. To make lighting the focal point of the room, use dramatic or sculptural fixtures that emit plenty of light, and place them at even intervals around a room. Examples of this type of setup include supplementing a spectacular pendant lamp or chandelier in a dining area with interesting sconces and using wall-mounted fixtures or wall washers on the perimeter of a living room and complementing them with torchieres that reflect light off the ceiling.

A lighting plan geared for plenty of bright light can quickly be transformed into one that also provides indirect or moody illumination with the addition of dimmer switches on all of the lighting sources in the room. For even more subtlety, forego the use of floor or table lamps. For out-and-out mystery, use hidden lighting sources to provide a soft blanket of background lighting in the room. These can be recessed in the ceiling or soffits or concealed behind beams, over or under shelving and cabinets, or behind large architectural elements that can be specially designed to accommodate them, such as cornices, moldings, baseboards, and coves.

left
Utilitarian light fixtures can provide great flexibility in meeting your lighting needs and at the same time complement your aesthetic tastes.

CHANGING SPACE WITH LIGHT Lighting can have a major impact on the depth and scale of a space. To make a room feel smaller, use downlights as the primary lighting source in a room and avoid lighting the perimeter of the space by positioning lighting toward the center of the room. This draws the eye into the room rather than out to its boundaries. To implement this, hang low pendant fixtures on the ceiling and make strategic use of lamps on tables and the floor.

To lower a high ceiling, keep light away from it by placing wall lights fairly low and using shades or pendant fixtures with closed tops that won't throw any light back on the ceiling. Also, draw attention to items placed at a low level, such as pictures or hangings positioned low on the wall or groups of accessories placed on low surfaces, by lighting them from above.

To make a room feel more open, airy, and spacious than it is, use indirect reflective light to create the illusion of height and depth. Wall washers trick your eye into seeing stretches rather than sharply delineating the boundaries of walls, while uplights aimed at the tops of walls visually increase their height.

right
Floor lanterns mark the way or become a piece of floor sculpture.

To make only the ceiling look higher, use floor or wall-mounted uplights to throw light up on the ceiling, or conceal lighting behind a cornice or coving mounted high on the walls at the perimeter of the room. To make a long, narrow space seem wider, focus attention on a feature at one of the end walls, such as a window with an elegant treatment or an interesting piece of art, by highlighting it with a spotlight, and wash the other walls in the room with an even but less intense light. To make a space seem larger, wash opposite walls with light to make them seem farther apart. Alternatively, combine lighting with mirrors and reflective surfaces to add the illusion of space to a room.

right

The lighting designer must create pools of artificial light to balance the natural sunlight flowing into the room.

CREATING MOODS WITH LIGHT There are myriad ways to use lighting to change the mood of a room. Some techniques focus on the color of the light, others address the decorative aspects of the fixtures, and others blend these approaches. In general, incandescent light makes a space feel warm, mellow, and cozy, thanks to the yellow tones it emits, while halogen and compact fluorescent lamps emit a cooler, whiter light that conveys a crisp, minimal, and modern aesthetic.

Fixtures in highly decorative shapes and styles lend a room their demeanor. For instance, for a rustic touch, use sconces, fixtures, lamps, even a chandelier made of a rugged material, such as wrought iron or wooden branches and boughs. For a romantic aura, choose antique or vintage fixtures, lamps, and sconces with delicate or curvy forms, and top them with pretty patterned or lacy shades. To imbue a room with a modernist or contemporary mood, choose lighting sources that are sleek and minimal, or consider concealing them with soffits, recessing them in ceilings, camouflaging them with architectural elements, or hiding them behind decorative elements such as pillars, pedestals, screens, and large plants or trees.

To make a space more intimate, use lots of table lamps to create a cozy glow and outfit them with shades in colors that are warm rather than cool, such as alabaster, pearl, parchment, or ivory instead of white, so the light they cast has a mellow tone.

__left
Use bedside accent lights for late-night reading, after the sun has set and the natural light no longer fills the room.

>**Quick** Fixes

Sometimes, changing a simple element in a room—such as the color of the walls or the strength of the bulbs in the fixtures and lamps-totally changes the nature and efficacy of its lighting. Here are quick and easy ways to correct common illumination problems:

> **Manipulate the amount** of light in a room with window treatments. Translucent drapes, curtains, and shades, which filter natural light and produce a diffused effect, come in many weights and should be chosen with regard to this property. Slatted blinds offer optimum control because they can block out all illumination in the room, flood it with light, or be used to strike a balance between these extremes.

> **If there are** several light sources in a room and the overall effect is still unsatisfactory, change the bulbs, which may be too bright or too dim for their surroundings. If a fixture or lamp emits too much glare, reduce it by replacing a standard bulb with a reflector bulb.

> **Change the shades** in a fixture to transform the magnitude and quality of the light it emits. Translucent materials allow more light into a room, which makes it seem brighter. White shades make light seem cool and clean, while shades with creamy hues emit more mellow tones. Darker shades can glow softly when the light is turned on and lend a space drama.

> **Use colored bulbs** to change the nature of the illumination in a space. Incandescent bulbs come in a variety of tints, and fluorescents can be covered with colored sleeves or manipulated with sheets of cellophane gel (which are clipped over the opening in a shade to tint the emitted light).

> **To make a** dark room much lighter, paint one or all of the walls a light color in a gloss or semi-gloss finish. Aim an accent light at one or two walls to reflect onto the other walls and brighten the room.

> **In a dark** corner, use a fixture with an opening at both top and bottom for more general illumination in the space. Make the corner a focal point by filling it with a tall plant or tree and positioning an uplight in front of or behind it. Experiment to find the most attractive placement.

> **If a room** is lit by a central pendant fixture that is inadequate, replace it with spotlights on a track system that runs through the central axis, or even criss-crosses it, for greater flexibility.

> **If the accent** lighting in a room isn't working, make it brighter. Accent lighting should be at least three times brighter than the general lighting in a room to highlight its intended target.

> **To instantly update** track lighting, outfit it with new fixtures or vary the fixtures on the track; they don't have to be the same and it can be both interesting and effective to mix spotlights and floodlights to create varying pools and points of light.

>**Minimum** Lighting Levels

While there are no hard-and-fast rules on how much light a specific space or room in a house needs, guidelines do exist. Keep in mind that many factors influence these suggestions, such as individual preference, the way a space is used, and the age of the individuals using the space.

While most people think of brightness in terms of wattage, professional lighting designers work in technical quantifiers such as lumens, foot-candles, and candela, using mathematical equations that are highly technical and site specific, which make them useless for the layperson. As a general rule, a room needs an average of 200 watts for every 50 square feet (4.5 square meters), and bulbs with higher lumens are more efficient given the same wattage (both lumens and wattage are listed on the package). Elderly people may need double the wattage because the pupil of the aging eye has less flexibility and eventually gets fixed in an open position that permanently demands more brightness. Average people in their fifties get as much light out of a 100-watt bulb as people in their twenties get from a 50-watt bulb.

To increase lighting in a room, either double the number of light sources or double the wattage of the bulbs, which is a more energy-efficient option if the fixtures can take it. However, keep in mind that wattage is only the beginning; the light a 100-watt bulb emits is affected by the efficiency of the fixture, the type of shade or covering over it, and its location in relation to where the light is needed.

Here are general guidelines for areas throughout the home:

> **Entry halls:** For an average 50- to 75-square-foot (4.5- to 6.75-square meter) space, use one 100-watt hanging fixture or 15-watt floodlamp in a recessed fixture.

> **Passageways:** Use one 75-watt fixture or one 100-watt recessed fixture for every 10 feet (3 meters) of hallway.

> **Closets:** For an average-sized closet, use one 100-watt fixture; for a walk-in closet, use one 100-watt recessed fixture every 10 feet (3 meters).

> **Living rooms:** Chandeliers or hanging pendants should have a low wattage and be used for decorative purposes. At a minimum, a small living room should have four table lamps or a combination of table and floor lamps. About 200 watts are needed on each wall so that a good base of background light, without harsh contrasts or deep shadows between background and task lighting, is created in the room. If a lamp is used to light an entire corner, it should carry at least 200 watts.

> **Dining rooms:** Chandeliers or hanging pendants are more important in dining rooms than in living rooms because they form a dramatic focal point over tables, but central fixtures that are too bright cause glare and visual discomfort and should be equipped with dimmers. Use low-wattage downlights (preferably recessed) on either side of a central hanging fixture for a low level of general illumination. Consider candles on the table for dramatic mood lighting. Candles should be high so diners do not look directly into the flames.

> **Bedrooms:** If a central lighting fixture is used for general illumination, it should be glare-proof (such as a shallow, frosted glass orb or disk) and about 200 watts. For reading in bed, use a table lamp no more than 25 inches (635 millimeters) away from the book or a swinging-arm wall lamp attached 12 inches (304,8 millimeters) from the bed, in line with the shoulder of the reader. Either should provide at least 100 watts of illumination, preferable on a three-way switch.

> **Dens, studies, and family rooms:** These rooms vary widely in size today, and it is necessary to approach large great rooms as a living room to obtain the correct amount of general background illumination. Table lamps and downlights are necessary to ensure enough lighting for activities such as reading, writing, playing games, and working on crafts and tasks. Three hundred watts is necessary for sewing, embroidering, or other highly detailed tasks. Downlights are also suitable for illuminating specific activity areas in the room; they should provide 250 to 300

watts to adequately light a whole area, such as a bar or game table.

> **Kitchens:** Ceiling fixtures provide a good overall level of shadowless illumination. These can be recessed or on tracks, but they should afford illumination to every single square foot of this space. Wattage needs vary according to the colors of the walls, tiles, and surface treatments in the room and the reflecting value of porcelain, tile, and metal surfaces. Avoid dark colors on countertops, as they absorb too much light, which can make certain tasks dangerous. Strip fluorescent lighting mounted under cabinets provides excellent illumination for the work surfaces below. To prevent shadows on the range and sink while working at these stations, make sure adequate lighting sources of about 200 watts are installed directly above them.

> **Washrooms and bathrooms:** High-wattage lighting is mandatory near the mirror behind the sink, but the specific level depends on the nature of the space. For instance, when two sinks are installed in a long counter, the length of the counter dictates the brightness of the lighting, which should be above or surrounding the mirrors behind the sink (or sinks). If the room has a freestanding sink topped with a medicine cabinet, use sconces or bracketed fixtures of at least 100 watts each on either side of the mirrored cabinet.

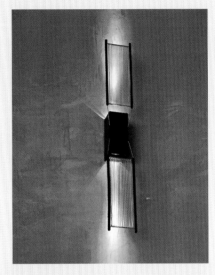

—**above**
Glass wings reflect the sconce's light vertically.

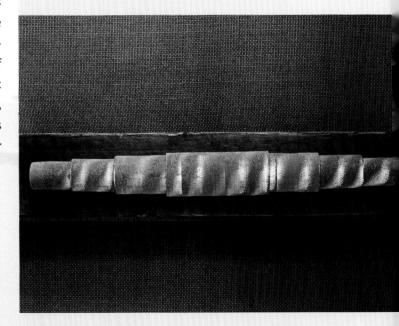

—**above**
Available in various shapes, sizes, and textures, a decorative sconce can set the mood for a hallway or room.

MAKING YOUR OWN LIGHT

A lampshade can be and should be much more than an unappreciated wallflower. It can act as an accent piece, soft-spoken and demure. It can punctuate its presence with elegant simplicity. Or it can step out as a bold statement, an intentional, unavoidable prop in a room's interior scenery. Just in its color alone a lampshade can influence whether a room feels warm or cool, subdued or vibrant.

A lampshade can reign in the harsh glare of a lit bulb without much—if any—attention to its size, color, or shape. It's the grim truth: a lampshade has the predestination to be drab.

However, hope for the neglected lampshade is anything but dim. What is beginning to transcend the long-forgotten shade is the arrival of a mind-boggling array of handmade and machine-made papers. You can readily find "paper" made from banana tree fibers, papyrus stems, or the hide of a goat. There are papers that look like crushed leather, wood grain, cloudy blue skies, cork facing. There are glorious wafer-thin lace papers, some with intricate patterns and others more organic, their patterns created by such elements as droplets of falling water. Many of these unique papers can be wrapped around or folded to make beautiful shades of all shapes and sizes. It can make night-lights, votive shades, and window covers, as well as lampshades. Light—be it from a basic bulb or candle's glow—becomes the perfect complement to paper's intricacies and marks of craftsmanship. With paper shades, the shade is not merely a means to insert light into darkness. The light, as well, comes to illuminate the shade and the awe-inspiring paper from which it is molded.

There's more good news: handmade and special machine-made papers can be remarkably affordable. With lampshade hardware being equally inexpensive, you can easily make your own lampshade, from some of the most gorgeous papers available, for under $20. Perhaps the most costly element to making a shade is your time, which all of the projects in this book set out to minimize by making the assembly as simple and straightforward as possible. Many projects are easily started and completed in an afternoon.

This book is divided into three categories: Traditional & Modern Shades, Exotic Shades, and Night-lights and Votives. So your choices are broad. Also, feel free to select a paper that best suits your personal taste. Each project offers general guidelines on the types of paper that will work with the shade. And most of the projects can be adjusted in size if you need to downsize or increase the size of the shade to better suit a lamp base.

CRAFTING TOOLS

Fortunately for the crafter, making lampshades does not require going out and purchasing a whole new array of tools. The basics will do. Listed on this page are most of the tools needed to complete the projects in this book. Not every tool shown here is required for every project. If any of the projects in this book require additional tools, they are listed in the specific project's materials list. For information on adhesives and glues see page 150.

TOOLS

PENCIL
A pencil and pencil sharpener are important elements when measurements are being made.

KNIFE
Utility or mat knives, X-acto knives, or knives with snap-off blades are all acceptable and easily available at hardware and art supply stores. A surgical scalpel and good-quality curved blades can be purchased through bookbinding or surgical suppliers. A utility knife is best for cutting binder's board.

SCISSORS
Make sure they're good and sharp. Used mostly for rough cutting. Most accurate cutting is done with a knife and straight-edge or, for the best of all worlds, heavy-duty board shears.

RULER
A metal ruler or straight-edge is preferable to a plastic one that can be nicked or shaved when used with a knife. The heavier the ruler, the more secure the action. A cork backing prevents the ruler from sliding while cutting is being done.

SQUARE RULE
A square rule is used with a knife for cutting boards and other materials, and maintaining right angles. It is available at an art supply store.

LARGE MAT BOARD
A self-healing mat board, imprinted with a grid pattern, on which endless cuts can be made with your knives, are available in several sizes from bookbinding and art supply stores.

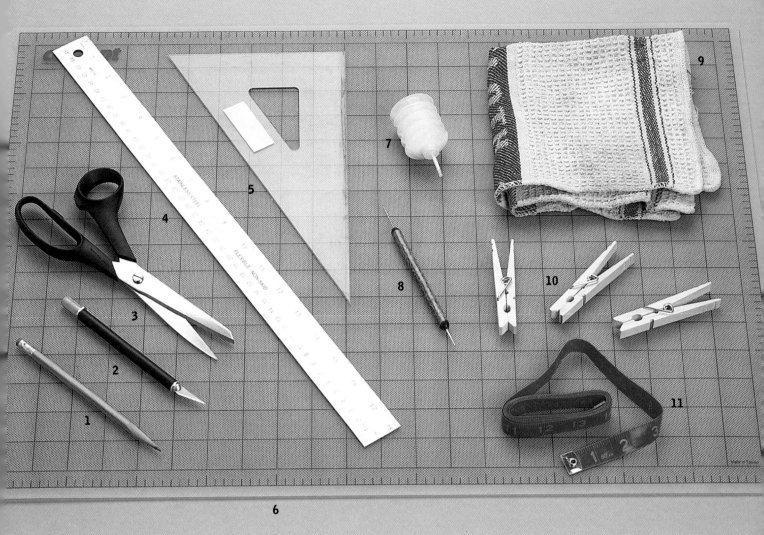

KEY

1. Pencil
2. Craft knife
3. Scissors
4. Ruler, preferably metal with a cork backing
5. Square rule
6. Large mat board
7. Glue dispenser with a long, narrow applicator
8. Awl
9. A damp rag for wiping away excess glue
10. Clothespins
11. Flexible seamstress measuring tape

GLUE DISPENSER
When choosing a glue dispenser, it's important to remember that a small hole lessens the potential for messy over-gluing.

AWL
A wooden-handled tool with a sharp, pointy metal shaft, used to punch holes. These are easily available at bookbinding suppliers and hardware stores.

DAMP RAG
Keep a damp cloth handy while gluing. Use it to clean sticky fingers so that you do not get any unwanted glue on finished surfaces.

CLOTHESPINS
Use clothespins to hold pieces in place while glue is drying.

SEAMSTRESS TAPE
For circular shapes where measuring with a ruler would be awkward, seamstress tape is invaluable. It is widely available at craft stores and sewing shops.

FIXTURES AND FITTINGS

Before you begin to buy any supplies and fixtures for your lampshade, it is important to take a close look at the hardware on your lamp base to understand the kind of lampshade fixtures for which it is designed.

The majority of lamp bases are designed so that the shade sets on top of a harp and is bolted in place with a finial. In such cases, the appropriate top wire fixture on a lampshade is a washer top wire (also known as a spider top wire). A shade made with a washer top wire permits the greatest flexibility in terms of the type of shade you construct. It can basically be used with nearly all shapes and sizes of shades. It does, however, hinge on your having a harp from which to hang the shade.

If your lamp does not come with a harp, but you wish to make a shade using a washer top wire, most hardware stores carry various sized harps. They are easy enough to clip right on. If your lamp needs a harp wing, however, to hold the harp in place, you may have to detach the socket, which requires familiarity with how a lamp is wired. In this case you may consider using a bulb clip adapter.

FIXTURES AND FITTINGS

BULB CLIP TOP WIRE
A bulb clip top wire is easy to pop on and off. It requires no specific hardware on the lamp base other than a bulb to which it can clip. A bulb clip top wire, however, does not secure to the lamp as tightly as a washer top wire, so such top wires are best suited to smaller shades that are less likely to shift when clipped to a bulb. If you already have a top wire that is something other than a bulb clip, you may still readily convert it to a bulb clip by adding a bulb clip shade adapter. If you wish, you can even purchase an adapter with a screw top to accommodate a finial. This comes in two styles: one for a standard bulb, the other for a flame-shaped bulb.

UNO TOP WIRE
This fixture actually props on the socket. Its inner circle is threaded so that it screws into place. Consequently, the lamp socket must also be threaded (known as a threaded uno socket) for the fixture to hold firmly in place. This style fixture is most commonly paired with floor lamps.

BOTTOM WIRES
No matter which top wire you choose, the bottom wire is always simply a plain wire with no special attachments. Its job is to attach to the paper arc or other form that makes the shade. With some lampshade frames the bottom wires come connected to the top wires as a complete unit. The side wires that connect the two are called "ribs".

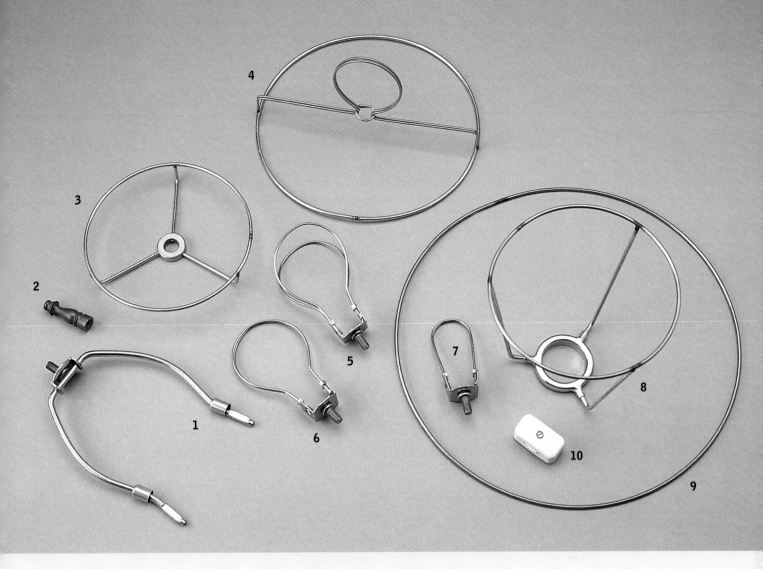

KEY

1. Harp
2. Finial
3. Washer top wire
4. Bulb Clip top wire
5. Bulb clip adapter
6. Bulb clip adapter with finial screw
7. Flame bulb clip adapter with finial
8. Uno top wire
9. Bottom wire
10. Cord switch

SHADE MEASUREMENTS

When ordering circular lampshade wires to make a drum-shaped, empire, or a cone-shaped shade, you will need the measurement of the shade's top and bottom diameter. That is the measurement that cuts across the center of the circular wire.

For most panel shades, be it a square, rectangular, or hexagonal shade, the shade frame comes in one unit. So you need the measurement of the top and bottom wires as well

as the side wires for height. In the case of a rectangle, that means you need two sets of three measurements since there will be two different sized panels on the shade. The square coolie shade on page 194 and the woven rectangular shade on page 180 are examples of single unit frames. You can also find square-shaped top and bottom wires sold individually. The basic box shade on page 162 is constructed with separate top and bottom square-shaped wires.

CORD SWITCH

For some of the projects in this book, the shape or construction of the lampshade requires that the power switch on the lamp be located on the cord. If your lamp base switches on at the socket, you can alter this by installing a cord switch, also known as an in-line switch. You may find these switches at a hardware store. They come with straightforward instructions for installation.

GLUES & GLUING

Perhaps the most intricate process in making shades is the gluing. Imperfections common to gluing—rippling, buckling, blistering—will show and quickly diminish the visual effect of your shade. While it is important that you take your time when gluing or adhering papers to make shades, it is also important that you use the right adhesive. Following is a list of the basic glues and adhesives you will find used in this book, along with a description of their qualities and uses.

TIPS FOR APPLYING GLUES AND ADHESIVES

- When gluing a paper lining to a shade wire, apply the glue to the wire not to the paper.

- Make sure to spread a thin, even coat of glue. Ideally, use a glue applicator with a long, narrow tip for more precise application. Cotton swabs are handy when you need to spread any type of glue.

- When gluing paper to a shade wire, start at the center edge of the paper and work out. This is especially important with a round shade. (See Step 1 on page 158.)

- Place clothespins close together when first gluing paper to a lampshade wire. This is to help avoid gaps. Once the whole shade has adhered flat against the wire, remove every other clothespin to let the glue dry.

- When adhering with double-sided adhesive tape, first stick the tape to the liner or base paper in the project. Expose the second side by removing the protective cover in increments. This way you have a better chance of making corrections if the paper is not properly aligned.

PVA-POLYVINYL ACETATE

This is the ideal glue for adhering paper to lampshade wires. It is white but dries clear. It does not dry stiff but instead is almost rubbery in consistency. It does not disfigure paper or cause blisters. Also, it has a tack fast enough to grab the paper quickly but slowly enough to allow time for you to make adjustments. When applying PVA glue, it is ideal to use a narrow tipped glue applicator. (See Crafting Tools, page 146.)

DOUBLE-SIDED ADHESIVE TAPE

Double-sided adhesive tape, readily available in craft stores, can and will save you a tremendous amount of time and frustration when assembling shades. This clear plastic tape with "stick" on both sides is ideal for making flawless seams on shades as well as adhering paper over shade liners. This tape is also effective when working with thin or fragile papers or constructing the votive projects in this book. This product is commonly found in strip/tape form but also comes in sheets. For the projects in this book, make sure the tape is designed for a permanent seal.

SPRAY ADHESIVE

A couple of projects call for this adhesive. Be sure to cover your work surface with newspaper or other scrap paper when using the spray. Its air-bound sticky particles travel. For a permanent bond, be sure to spray both sheets of paper that are to be glued together.

NEEDLE AND THREAD

In some projects here, the paper is not held to the shade with any glue or adhesive but, instead, with stitching. This technique is primarily used with delicate, fabric-like papers that are gathered or attached to shade wires without a lining.

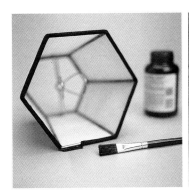

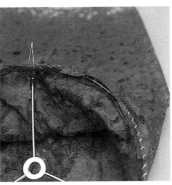

SURFACE EFFECTS

While you might start out with a flat sheet of paper, there are a wide variety of techniques you can apply to paper to create surface effects to your shade. These can enrich your shade with texture and add interesting play with shadows and light.

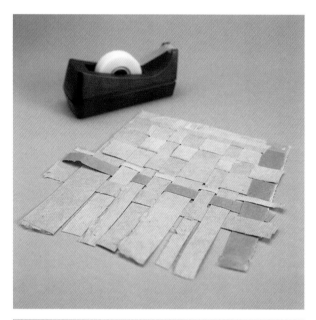

One way to change the surface of the paper is by manipulating it in some way. Crumpling, folding, gathering, overlapping, and fluting are all ways of creating varying effects. Weaving is another technique for creating surface texture.

Two other popular techniques for affecting the surface of a shade are cutting and piercing, both applied to projects found in the votive chapter starting on page 218. Cutting and piercing both offer a glimpse behind the shade, much like catching a peek inside a lit home. The effect is always intriguing. Placing these shades over the flicker of candlelight adds all the more visual enticement as the illuminated patterns dance against a dark wall.

You can easily experiment with both piercing and cut outs to formulate your own patterns and designs. Your designs can be geometric, abstract, or an outline of a favorite form or shape. You can also combine both techniques in creating your design.

It is always wise to place a protective covering over your work surface, such as a thick towel or a few layers of felt, when you are piercing paper. An awl (see Tools, page 146) is commonly used for piercing, but just about any sharp, pointed tool you can find will do. Various sized needles, an ice pick, or even a sharp nail would suffice. You can also play around with various sized piercings. Perhaps your largest might not even be a piercing at all but a circle created with a hole punch. Experiment.

With cutouts, it is easy to custom design your own pattern. The key is to avoid connecting lines in the paper. Similarly, when piercing, distance the holes by at least 1/8" to avoid breakage. It is best to create a template before cutting or piercing into the paper that is to be used for the finished shade. If the pattern is to be continuous around a shade, it is often a good idea to sketch a grid on the back of the paper you are working with as a guide for the pattern's even placement.

An effective finishing touch to pierced or cutout paper shades (or a flattering combination of both) is to back the shade with a contrasting translucent paper. This will help to enhance the design you've created and add an additional dimension of color to the shade.

THE BASE-ICS

When making a lampshade, there's no such question as "which came first: the chicken or the egg?" Before anything, you need to identify the base you are to use—unless you are making a hanging lamp. The good news is that a lot of the conventional rules for matching bases and lampshades have fallen by the wayside. So what you primarily need to do is to rely on your solid intuition, creative flair, and personal touch. Your lamp base and shade are meant to be an ensemble. It is almost like putting together an outfit. You will want both the components to complement one another—not to compete with each other for attention.

Dimensionally, the shade should reflect that of the base. Think balance. An itsy shade atop a hefty base is sure to look awkward. That is not to say, however, that you cannot try pushing the envelope. The idea is not necessarily to make the lamp invisible, camouflaged into the backdrop. It is to highlight its attributes and individuality. So while it is intuitive to coordinate a square base with a paneled shade, such as a square or rectangle, a drum shade could add an unexpected, pleasant flair.

Whatever shape or size you choose, do make sure your shade is long enough to cover the socket but not so long that it is hovering over the base. If you have either of these problems, one simple solution might be to change the size of your harp. A smaller harp will drop the shade; a larger one will give it life. If you are making a round empire shade and are uncertain about what length the sides of your shade should be, you can experiment by simply tying a few strings to suspend the bottom wire from the top.

ABOUT PAPER

Once you have decided on your lamp base, you can start to consider your paper—almost. Your first step in selecting a paper is to determine the function the lamp is to play in the room. Will the lamp be used for general lighting? Task lighting? Or will it simply function as an accent piece? The ramifications of your decision are two-fold. With your selected paper, you are creating a visual component—a shade—to the decorative landscape of a room.

Furthermore, that very shade will create a visual environment of its own by the manner in which it sheds light. Think about how far you wish your lamp to cast its light. Do you want the light to be focused or to splay across the room? Or maybe you want it to cast its light towards the ceiling. Also, what kind of feel do you wish the shade to create? Are you looking for warmth, or something modern and cool? It is the very paper from which the shade is constructed, not just the shape of the shade that creates an effect.

When selecting a paper, consider the paper's qualities—its strength, flexibility, and translucence. All can be crucial to the construction of your shade. For example, you don't want to make a gathered paper shade with a heavyweight, rigid paper. Rather, you need something soft and flexible, almost fabric-like. To optimize brightness, you will want to avoid a dark, opaque paper, which is more apt to create a somber atmosphere. Instead, choose a paper that is light and highly translucent. All of the projects in this book take these factors into account and offer suggestions for alternate papers if you cannot find the specific paper called for in the project.

As you are considering a particular paper for a shade, be sure to hold it up to the light. This helps you to gauge its translucency and also predict the characteristics of the paper that a light will enhance. Many papers take on a magnificently richer appearance when set before a light. Depending on your project, this may be desirable—or it may not. Similarly, feel the paper. Get a sense of its surface and substance. Many paper stores have binders filled with samples of their papers so that you are at liberty to touch them as much as you wish.

When shopping for your paper, be sure to bring a notepad with the measurements you need. If you can, also bring the lamp base. If in doubt, most paper stores will supply you with swatches at no cost. Take them home and consider your options. If you are ordering paper by mail, it is a particularly good idea to first request paper samples.

In addition to choosing a paper for the exterior of your shade, you will find that most of the shade projects in this book call for a lining. Heavyweight vellum or parchment papers are most often recommended since they are both translucent and sturdy in nature. The vellum does tend to be slightly more translucent and uniform. Parchment can be mottled but can also be easier to work with.

Finally, before making the ultimate commitment—cutting into the paper of choice—it is always good to make a mock-up of your shade with scrap paper. It helps you get acquainted with the shade-making process. It can also save you from making mistakes and wasting money on your prized paper of choice.

LAMPSHADE CRAFTING 101:
The Basic Shade

The round empire lampshade is the most basic of shades. Knowing how to make this classic is helpful in understanding many of the fundamentals behind making most any lampshade.

The following is a step-by-step outline of how to make a basic empire shade. Included are tips that will help you avoid potential problems in the process of crafting this and many other kinds of shades. Therefore, it is recommended that you read this section, and the following one on trims, before proceeding with any of the projects in this book.

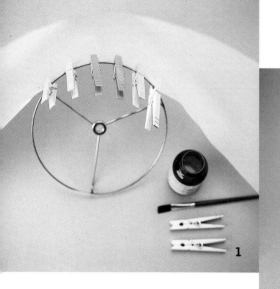

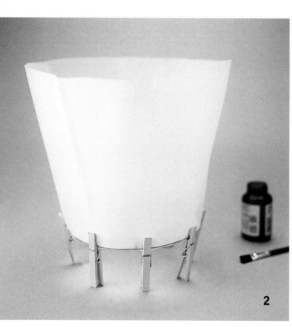

Step 1

Glue the arc-shaped lining paper, cut from a template, around the outside of the top wire. (See Templates page 278) Apply glue in a thin line, spreading evenly with a cotton swab if necessary. Start gluing at the center of the arc and work out, all the while securing it with clothespins placed close together to prevent the paper from shifting or gapping.

Tip: Unless specified, it is important with most paper shades to start out with a lining. Heavyweight vellum or parchment paper is most often recommended since they are both translucent and sturdy.

Step 2

Once the arc is attached to the top wire, remove every other clothespin and let dry.

Tip: Often the best way to prop the shade up so that it does not warp or shift while drying is to place it upside down so that it stands on the end tips of the clothespins.

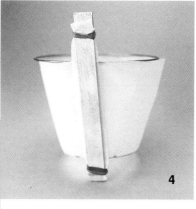

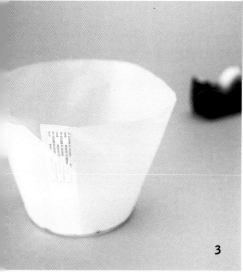

3

Step 3

Adhere the sides of the paper arc, which should overlap about 1/2", to form a seam. Double-sided adhesive tape is ideal.

Tip: Before sealing the seams, tape them together at the bottom of the shade first and insert the bottom wire to make sure it fits properly.

Step 4

If using glue to seal the seam, the following will insure that it dries without puckering. Sandwich the glued seam between 2 strips of foam package padding. Press the padding firmly against the seam by covering with two rulers or wooden slats held together with elastic bands.

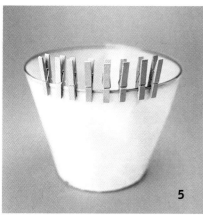

5

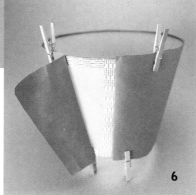

6

Step 5

Set the shade upside down, and glue the bottom wire to the bottom of the shade. As in step 1, start from the center of the arc and place clothespins close together while attaching to prevent shifting and gaps. Once the shade is attached to the wire, remove every other clothespin and let dry.

Tip: Vellum tends to pucker as the glue dries. To smooth any rippling or puckering where the vellum has been glued to the frame wires, blow on it with hot air from a hairdryer. The heat will help to pull the paper taut.

Step 6

Apply double-sided tape or a line of glue around the top and bottom edges of the shade, as well as along the seam. Wrap around it the shade paper, which should have been cut into the same arc shape as the lining.

Now your shade is ready for the finishing touches.

TRIMMING LAMPSHADES

Besides giving a shade a clean, finished look, trim can be important in helping to hold the shade together. If you want to finish your shade with a traditional trim, it is important to make sure you select the correct type and size ribbon. Many ribbons, such as satin, will react negatively to glue and pucker and slide out of place. The ideal ribbon for trimming a paper shade is a ribbed ribbon known as grosgrain, which is made of a combination of 43 percent cotton and 57 percent rayon. Typically, those companies that sell lampshade fixtures also sell this particular trim in a wide assortment of colors and a variety of widths. To determine the size trim you need for your lampshade, measure how wide a trim outline you wish to have around the shade. Add that to the extra width you need to have the trim wrap over the frame wires and inside the shade. To measure the length of the trim, you need to measure the circumference of both the top and bottom of the shade. Use a flexible seamstress's tape to do so.

The following is a series of basic steps for applying trim. For hints on how to work with trim when applying to angled shades, see the trim steps for the hexagonal sconce shade on page 187.

1. Before applying the trim, use a ruler and pencil to mark where the edge of the trim should fall along the shade's top and bottom edges. This marking gives you a guide to follow when gluing on the trim so that it falls in a straight line. Allow the trim glued to the outside of the shade to dry before adhering the remainder to the inside.

   Tip: When gluing trim to a shade, apply a thin layer of glue to the trim, not to the shade, and do so in 3" intervals as you work around the shade.

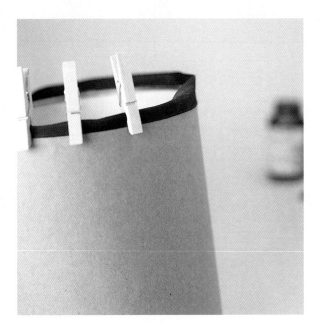

2. Again, working in intervals, spread glue on the remaining trim and wrap it inside the shade, tucking it under the wires. Hold it in place with clothespins to dry.

3. When you run across the spokes of a frame's top wire, clip the ribbon to make a "V" so that the trim accommodates the spoke.

4. Always cut the ends of trim at a diagonal to prevent the ribbon from fraying. To complete the trim, glue the diagonal ends so that they overlap. When applying a narrow trim, such as the hemp cord here, finish by gluing the ends so that the tips meet.

5. Depending on the shade and the trim used (if any), you can also employ the end of the trim as an accent to the shade. Here the ends of the hemp cord were simply knotted together. Other ideas might be a button or a lightweight buckle.

BASIC BOX SHADE

The simple, clean lines of a box-shaped shade translate into a confidently understated form that needs few or no embellishments. For this project, a machine-made paper from Thailand known as unryu was used. This paper is made from mulberry pulp into which fiber strands are embedded. The effect is a lightweight, translucent paper that is ethereal yet deceptively strong. It seems fitting that unryu literally means "cloud dragon paper." This soft paper is not known for its folding properties, so it is wrapped around a stiff folded vellum box structure. This process ensures that the unryu fits snugly around the crisp corners of the shade.

The construction of this shade is straightforward. Instead of securing the paper to the wire framework with trim, the top and bottom edges of the paper fold at 90-degree angles toward the center to create a lidded effect. This serves two functions: it underscores the box feel without obstructing the passage of light, and it creates a structural flap so that the shade actually sits on the top frame wire, simplifying the construction.

Paper Choices

The options for paper are quite broad with this shade. The shape of the shade works best with a paper through which light can be emitted (otherwise, the light can seem too "boxed" in). The paper should also be able to make clean folded edges. Or, like the unryu, it should be supple enough that it can wrap snugly around the edges of the vellum so that there is still a crisp edge to the box shape.

Materials

- 1 strip of 25" x 7" (64 cm x 18 cm) plain, translucent vellum

- 1 strip of 25" x 7" (64 cm x 18 cm) orange unryu paper

- 1 square washer top wire, 6" x 6" (15 cm x 15 cm)

- 1 square bottom wire, 6" x 6" (15 cm x 15 cm)

- 1 finial

- Basic craft supplies (see page 146)

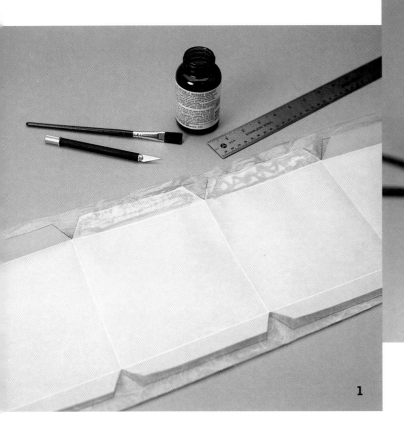

1

Step 1

Use the template on page 279 to cut the vellum. To make folds, use the blunt side of a craft knife to score the vellum where the dotted lines on the template are located. Fold along the scored lines in the same direction. Fit the folded vellum cutout over the unryu paper with the flaps folding away from the unryu paper (as shown). Using double-sided tape or a thin layer of glue, adhere the vellum to the unryu paper by its flaps starting with the top and bottom center flaps and working out. The four, 6" (15 cm) squares in the center of the vellum do not need to be glued to the unryu paper but should lie flat against it.

Tip: Vellum is likely to wrinkle when it comes in contact with glue. To minimize this effect, spread the glue onto the unryu paper, not the vellum. Use a light, even coat of glue. Let dry slightly, about 20 seconds, before adhering the vellum to the unryu paper.

Step 2

Use a craft knife to trim the unryu paper around the flap edges so that it fits flush with the edges of the vellum cutout. Apply double-sided tape or spread a thin, even layer of glue to the unryu paper along the inside (the vellum side) of the left side flap. Fold it over so that it sticks to the inside edge of the adjoining 6" (15 cm) vellum square. Run an even line of glue along the outside (the unryu side) of the right side flap. Bring the right side flap over to the glued left flap and adhere so that the unryu sides of the flaps are adhered together forming a box shape (as shown). Be sure the corners of the flaps are perfectly aligned for a nice square box shape. Secure with clothespins and let dry.

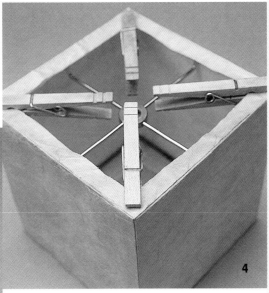

4

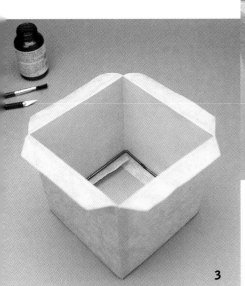

3

Step 3

Prop the box shape upright, folding the four, 1" (3 cm) wide bottom flaps under. Run a line of glue on the inside of these flaps along the crease line. Insert the bottom frame wire inside the box so that it sits on the lines of glue. Let dry. To glue the top wire, open the top flaps—so that the vellum side faces out—and run a line of glue along the crease line. Insert the top wire into the box. While holding the washer top wire in place inside the shade, fold the flaps back over so that the glue comes in contact with the wire. Carefully flip the box upside-down to let dry.

Step 4

To finish, glue the top flaps together where they overlap in the corners. Secure with clothespins and let dry. Then glue the bottom flaps in the same fashion.

Troubleshooting

To make sure the lightweight bottom wire adheres snugly to the bottom flap edges as it dries in step 3, you may wish to set something on top of the wire as it dries. An easy solution is to place the top frame wire on top of the bottom wire. Then place something of weight, such as a can of beans, on top of the washer frame, where the finial would sit, to anchor the bottom wire onto the paper flap. The top frame wire may also need to be weighted down in a similar fashion.

Variation

Using a striped pattern for the box-shaped shade will make a bolder statement than the soft fibrous unryu. For a more finished appearance, you might wish to place trim around the folded edges or on the inside edges of the top and bottom flaps.

MODERN AND TRADITIONAL PAPER SHADES

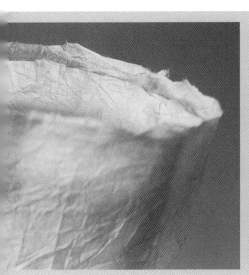

PAPER TWIST CONE

This project uses readily available materials and supplies. The slender conical shade is decorated with everyday twisted craft paper, often used for bows on wreaths or floral arrangements. It is readily available at craft supply stores as well as at many general housewares stores. A liquid laminate will give the shade a glossy finish—a modern, vinyl-wrapped look. The tall, narrow shape of this shade makes it a perfect choice for a table where space is limited. Its lack of transparency makes it best suited as an accent lamp. This way the light bulb sits lower and at a safe distance from the sides of the shade. Be sure to use a tapered light bulb, such as those shaped like a torch flame.

Materials

- 1 sheet of 14" x 20" (36 cm x 51 cm) dark, rigid paper or poster board (navy or black)

- 1 string of 6" (15 cm) marine blue twisted craft paper

- 1 string of 16" (41 cm) marine blue twisted craft paper

- 1 string of 20" (51 cm) marine blue twisted craft paper

- 1 string of 22" (56 cm) marine blue twisted craft paper

- 1 string of 15" (38 cm) marine blue twisted craft paper

- 1 washer top wire, 4" (10 cm)

- Liquid laminate

- Basic craft supplies (see page 146)

1

2

Step 1

Unravel the measured twisted craft paper strings loosely, flattening them without pressing out the fine creases. The strings should unravel to a width of 2 $\frac{1}{2}$" to 3" (6 cm to 8 cm).

Step 2

Use the template on page 280 to cut the dark paper or poster board into a cone shape. Glue the unraveled paper strips to the cone at a diagonal. Begin by gluing the 22" (56 cm) strip. Its lower left edge should sit at the lower left corner of the cutout. Its upper right edge should lay 2" (5 cm) below the upper right hand corner of the cutout (as shown). Lay the remaining strips at the same angle. The strips should be placed from the top right hand corner to the lower left hand corner in the same order listed above under "Materials." All of the strips hang over the cutout's edge by at least $\frac{1}{2}$" (1 cm).

Step 3

Trim the glued strips so that there is a $\frac{1}{2}$" (1 cm) overhang, except on one of the straight edges, where you should trim the strips flush to the straight edge.

Tip: Because of the creases in the twisted paper, you may find it easier to use scissors instead of a craft knife when trimming the untwisted strips of paper.

This shade is designed to be rigid since the frame sits in the center of the shade and does not provide much structure. If you are using poster board to make a rigid cone, it may be too stiff to hold in a cone shape to glue the seams together in step 5. If so, let the shade acclimate to the shape overnight. Roll the shade into its cone shape. Hold the ends of the cone into shape by clipping the seam ends with small spring-loaded binder clips. Place a good padding of paper towel or tissue paper in between the clips and the shade so that they do not leave an imprint in the paper. Tie 3 strings about 2" (5 cm) apart around the center of the cone to cinch it in place. Let the shade sit overnight in the cone shape. To glue the seams together, remove the spring-loaded clips but not the strings. Work around the strings, which help to hold the cone shape in place as the glue dries.

Step 4

Along the top curved edge of the cutout, make slits every $\frac{1}{2}$" (1 cm) along the overhanging marine strips, cutting to the cone edge. Do the same along the bottom curve at 1" (3 cm) intervals. Glue the overhanging marine strips along the 3 edges to the cutout.

Step 5

Shape the cutout into a cone shape overlapping the straight edges by $\frac{1}{2}$." Glue the overlapping straight edges together, placing the edge with the marine paper folded over it on the outside (see troubleshooting). Let dry. Coat the outer edge of the wire frame with glue and place inside the cone as deep as it will fit (about 3" (8 cm) from the very top). Be sure it is set straight in the cone; otherwise, the shade will sit on the lamp lopsided.

Variation

Indulge in a bit of creativity by experimenting with the concept of using twisted craft paper to decorate lampshades. For example, instead of untwisting an entire strand, in this variation the paper was unraveled only at intervals.

FLUTED DRUM SHADE

When patterned paper is fluted, a certain degree of intrigue is created as the pattern travels in and out of the shadows of the shade's curves. Linear prints are particularly effective with this style of shade. Here a gold print was used to complement the base. A thin gold soutache trim follows the wavy paper edges to give the shade a more finished look. For a more modern look you might use a less controlled patterned paper or you could reverse the approach. That is, use a bold patterned paper for the shade lining instead of the solid green used in this project. Then use transparent vellum for the fluting. Whichever approach you take, the key to making a fluted shade is being compulsive about accuracy. The design is easily adjusted to varying sizes of drum-shaped shades as well as to the width of the paper's flute. To do so, see instructions in "Variations" on page 175.

Paper Choices

Knowing the size of the flute is key when selecting a paper for any fluted shade. Knowing the size, you can test whether the paper can comfortably be fluted. For any fluted shade, a paper of medium weight is key. It must be supple enough to bend with the curves without buckling, yet it must be sturdy enough to hold its form without sagging.

Materials

- 1 sheet of transparent vellum, cut into a 7 ½" x 26" (19 cm x 66 cm) rectangle

- 1 sheet of 8 ½" x 26" (22 cm x 66 cm) medium green unryu paper (or other background paper)

- 1 sheet of 7 ½" x 40" (19 cm x 102 cm) India Woodblock paper (or other patterned paper of soft, medium construction weight*)

- 8" (20 cm) top lampshade wire

- 8" (20 cm) bottom lampshade wire

- 2 yards (1.8 m) plus 1 foot (.3 m) of ⅛"-inch (.3 cm) gold trim

- Basic Craft Supplies (see page 146)

*If you cannot find a sheet of paper 40 inches long (102 cm), then glue two sheets together. However, be sure that the overlapping seam falls on the inside curve of the fluting flush against the shade lining. (See step 4.)

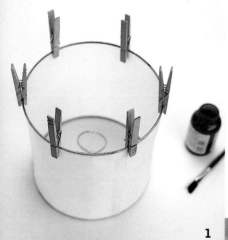

1

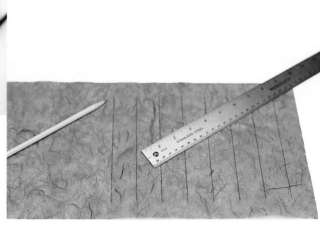

2

Step 1

Run a line of glue around the top shade wire and attach a long edge of the vellum sheet. Use clothespins to hold it in place as the glue dries. Before gluing the seam, fit the bottom wire, holding it snuggly in place with clothespins. This way you will know if the seam needs minor adjustment. Adhere the seam using double-stick tape or glue. If gluing, let dry. Then, as with the top wire, glue the bottom wire to the opposite long edge of the vellum sheet to form a drum.

Step 2

Lay the green paper flat on your work surface. Pencil in the following grid as a guide for the fluting placement: 1" (3 cm) below the top, long edge of the paper and 1 ¼" (3.5 cm) in from the left edge pencil in a 5" (13 cm) vertical line. Repeat across the paper, spacing the lines 1 ¼" (3.5 cm) apart. Or, if your paper is too dark for the pencil to show through, use a white colored pencil or fabric pencil.

Step 3

Wrap the green paper around the drum with the grid facing out, allowing a ½" (1 cm) overlap on the top and bottom edges. Run a line of glue along the inside edge of the top wire and about ¼" (.5 cm) beneath it as well. Pull the green paper snuggly over the glue-coated top wire and press against the inside edge of the vellum drum where the second line of glue was run. Press down so that the paper is flush against the wire and drum interior. Repeat the gluing process at the bottom of the shade clipping with clothespins to hold in place if necessary.

4

Step 4

Place the patterned paper flat on your work surface with its patterned side (front side) facing down. On the backside, draw a 7 ½" (19 cm) vertical line down the paper, two inches (5 cm) to the right of the left short side of the paper. Repeat across the paper, spacing the lines 2" (5 cm) apart. Make a gentle fold along each penciled line, folding so that the front side of the paper is on the inside of the fold.

Step 5

With the patterned side of the paper facing out, align a seam edge of the patterned paper with the seam edge of the green paper drum. Run a thin line of glue along the green paper's seam edge and cover with the edge of the patterned paper. Clip in place with clothespins to dry. Next to the seam in the direction in which the patterned paper is wrapping, run a thin line of glue along the penciled vertical line drawn on the green paper.

5

Align the first fold line of the patterned paper with the line of glue on the green paper to form the first flute. Clip to the side of the flute to hold it in place as it dries. Continue gluing the fold lines every 1 ¼" (3.5 cm) along the drum's circumference until you have fully worked around the shade. Finish by gluing the gold trim around the top and bottom edges of the fluted shade.

COFFEE-DYED LAMPSHADE
with Copper Lashing

Crumpling soft, Koji paper creates a shade with an organic elegance. Staining it in coffee lends an additional vintage feel. By lashing the soft, cylindrical shape with copper, the shade is pulled together both literally and figuratively. The effect is an hourglass-shape table lamp that channels light upwards and downwards, and defines space as it accents a room. The upward dispersal of light helps to further draw the viewer's attention to this lovely piece.

To make this shade you need a "slip uno frame" (see page 148 for more). The upside-down cone shape of the frame will fit snugly on the lamp socket. Two cones made from stiff vellum are joined at their smallest ends to form an hourglass shape: They basically hang from the uno frame and provide a framework for the soft crumpled paper to fit around.

Important note: *The unusual shape of this shade makes it difficult to reach the socket to switch the light off. If the lamp base you are working with does not have a cord switch, you can buy a cord switch at your local hardware store and easily install it yourself. (See Fixtures and Fittings, p. 148)*

Paper Choices

This shade calls for paper that is supple enough to be crumpled without marked resistance—but strong enough to stand on its own (without framing or structural wires) after it is crumpled and shaped into a cylinder. If you choose to dye the paper, you must also be certain that it is absorbent. Koji, a native paper of Japan, is an ideal choice. To create a more daring look, another option would be a foil-type paper.

Materials

- one large pot of strong, warm coffee
- two old bath towels
- two medium-size elastic bands
- two 16" x 22" (41 cm x 56 cm) sheets of Koji paper
- two 12" x 19" (30 cm x 48 cm) sheets of stiff, translucent vellum
- one 5" x 4 1/2" (13 cm x 11 cm) slip uno lampshade frame
- 30" (76 cm) string
- 34" (86 cm) copper wire
- Basic craft supplies (see page 146)

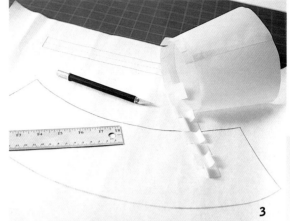

3

1

Step 1

Crumple each sheet of koji paper into a relatively tight ball. Secure each ball with two rubber bands to hold in place. Pour the warm coffee into a medium heat-proof bowl or saucepan. Submerge the paper balls in the coffee and let sit for about one hour so that they become fully saturated. For an even darker shade, let sit an additional hour or more.

2

Step 2

Remove the balls from the coffee and gently unravel, taking care not to tear the paper. Loosely spread each sheet out flat on a towel to dry. Don't smooth the paper too much, since part of the effect is the way the dye collects and dries in the creases. Be sure to use towels that you use around the house for rags since the coffee will stain them.

Step 3

Use the templates on page 282 to cut out two cones and a long narrow "collar" strip from the vellum. Score the center of the collar strip with a craft knife, using the dotted line as a guide. Next, make slits about 1/2" (1 cm) apart from the score to the long edge of the collar. Repeat on the other side, staggering the slits to make "flags," as shown above. Glue the sides of each cone template to form two equal-sized cones 1/2" (1 cm) overlap at seam. Then glue the short ends of the collar strip together 1/2" (1 cm) overlap to form a band.

Step 4

Create an hourglass shape with the cones by gluing the top half of the band to the in-side of the cone at its smallest opening (bend back the slit "flags" so that they adhere to the cone's interior). Once dry, glue the bottom half of the band to the small opening of the second cone in the same manner. Brush glue on the top and side wires of the uno frame and slip it into one of the cones, as shown. Clip around the edge of the cone to hold the frame flush against it while the glue dries.

5

Step 5

Taking a sheet of the stained, crumpled paper, overlap the 16" (41 cm) edges by ¹/₂" (1 cm) and glue together to form a cylinder. Repeat with the second sheet of paper. Allow both to dry thoroughly, then slip one inside the other. For this shade, a light, relatively smooth koji was selected. Its weight and texture are easy to work with, and would be best likened to that of cotton tissue paper or a sheet of fabric softener.

6

Step 6

Set the hourglass cone on top of the lamp socket so the uno frame is on the top half. Slip the double-layered cylinder over the cone. Making sure it's centered, grasp the paper cylinder at the "waistline" of the hourglass and secure with string. Adjust the cinched crumpled cylinder so it gathers and softly flares. Wrap the copper wire around the "waistline" and twist a few times. Curl each of the wire ends in to make small loops.

Variation

Try fabric dye for a brighter color range. For a uniform beige tone, soak the paper in coffee without crumpling it first, smooth flat to dry. Instead of coffee, stain absorbent paper with tea, the juice of crushed berries, beet juice, or fabric dye. Test your dye source on a small swatch of your paper. Another variation is to twist the paper instead of crumpling it to get creases that run vertically along the shade, as was done with both of these pieces. Here, the solid colored swatch was dipped in the dye first and then twisted for an even coloring. The mottled "tie-dye" effect of the second swatch was created by twisting the paper, securing it so it did not unravel, and then soaking it briefly in the dye so it did not become fully saturated.

BASKET WEAVE

Crafting this shade can carry you back to grade-school days of weaving paper simply for the amazement of it—or for crafting paper Easter baskets for the joy of it. You can take a wide variety of approaches with this shade. For example, you can tear the strips for a more natural look; or cut them with a craft knife for more precise, clean lines. You can cut strips of even width for uniformity in the weave; or vary them for a slightly more modern, personal touch. The look of the shade is equally as affected by the color scheme as by the weave pattern, i.e. how you place the different colored paper strips in the weave. Feel free to play with this shade. It easily adapts to a square frame. It also can be made to wrap around a drum-shaped shade. A round or empire shade is the one basic form that does not translate easily to a basic weave.

Materials

- 1 rectangular lampshade frame 12" x 10" x 8" (30cm x 25cm x 20cm) and 7" x 5" x 8" (17.5cm x 13cm x 20cm): (measurements indicate length of bottom wire by length of top wire by height)

- 1 sheet of 20" x 30" (50cm x 75cm) neutral colored vellum or parchment

- 1 sheet of 20" x 30" (50cm x 75cm) dark sand-colored handmade paper

- 1 sheet of 20" x 30" (50cm x 75cm) pale peach handmade paper

- 1 sheet of 20" x 30" (50cm x 75cm) pumpkin-colored handmade paper

- 3 yards (2.7m) of ½" (1 cm) wide sand-colored paper trim

- Scrap paper

- Adhesive tape

- Basic craft supplies (see page 146)

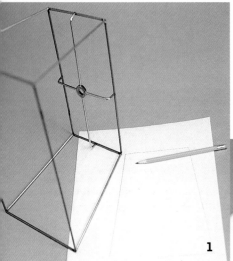

1

Step 1

Trace each side of the frame onto the neutral colored paper. Cut out each tracing and glue it to the wire frame to create a backing for the woven shade.

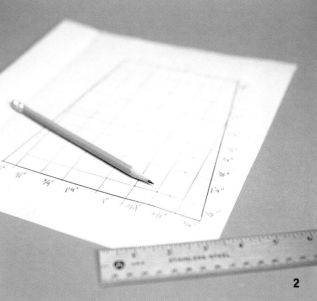

2

Step 2

On a scrap piece of paper, trace one of the long and one of the short sides of the frame. Using pencil and ruler, draw a mock grid of how you wish the woven pattern to look. You can go with varied sized strips for a contemporary look or strips of the same width for a more traditional basket weave. Mark the width of the strips needed along the side of the grid.

Step 3

Tear vertical strips from the four different colored hand-made papers according to the grid you designed. There is no prescription for how many strips of a particular color need to be cut. This is up to you. For example, if you wish the shade to take on a stronger peach/pumpkin tone, tear more strips of those colors. Align the vertical strips next to each other on a piece of scrap paper and secure in place at the top with tape.

4

Troubleshooting

Before gluing the frame to a woven swatch, make sure the weave is square. That is, make sure the vertical and horizontal strips are intersecting at 90-degree angles. Also, adjust large gaps between strips. Slight gaps, however, can make for a nice effect when the bulb is lit and slits of light shine through.

Step 4

Tear the horizontal strips of paper according to the grid To weave, start with the top horizontal strip and run it over and under the vertical strips. Continue down the grid, alternating between "over" and "under" for the starting point of each horizontal piece.

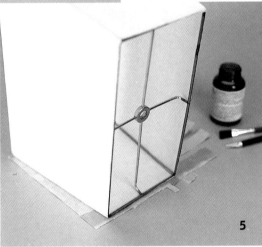

5

Step 5

Lay the frame over the woven pattern making sure it is square with the weave, i.e. the edges of the top and bottom horizontal strips are in line with the top and bottom frame wires. Glue around the frame edge and set the frame on top of the woven pattern to adhere. Once dry, trim away the portion of the weave that hangs over the frame edge. Repeat, gluing weave to the other three sides of the frame and trimming away the excess. Once complete, apply sand-colored trim around the top and bottom edges of the shade, as well as down the side corners. This trim should only run on the shade front; it is apt to split if folded over the wire framework.

Variation

Instead of tearing strips of paper, save time by weaving strips of paper ribbon.

HEXAGONAL SCONCE SHADE

Capping a sconce lamp fixture with a hexagonal shade adds an elegant touch to the wall of any room or hallway. This shade is designed to fit over a narrow flame-shaped bulb. For a larger lampshade, follow the same instructions for this project, but adjust the measurements of the paper and trim accordingly. The same general instructions can also be easily applied to making "paneled" shades of other shapes.

Contrasting trim on a panel shade will provide a sharp border. If you wish to avoid this look, choose a trim that is closer to the color of the shade. Always use a narrower trim down the side wires of the lampshade frame than that along the top and bottom wires.

Materials

- 1 sheet of 20" x 25 $\frac{1}{2}$" (50 cm x 65 cm) white crinkled paper

- 1 large sheet or 6 sheets of 8 $\frac{1}{2}$" x 11" (22 cm x 28 cm) white vellum

- 3 feet (2.7 m) of $\frac{3}{8}$" (1 cm) black grosgrain trim

- 3 feet (2.7 m) of $\frac{3}{8}$" (.5 cm) silver soutache trim

- 1 clip hex frame, 4" x 6" x 5" (10 cm x 15 cm x 13 cm), for a flame-style bulb

- Wooden clothespins

- Basic craft supplies (see page 146)

Step 1

On a scrap piece of paper, outline one side of the shade to make a template. Use the template to cut out 6 vellum panels. (See Troubleshooting and page 283.) Use the same template to cut out six panels of the crinkled paper.

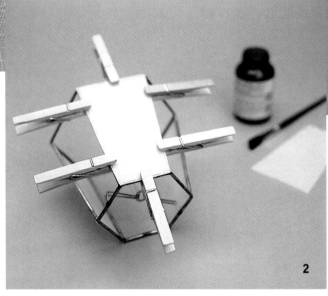

Step 2

Apply a thin line of glue along the wires surrounding one panel of the frame. Center one of the six vellum cutouts over the panel wires. Beginning at the center and working out, use clothespins to clip the paper first to the top and then to the bottom wires. Next, clip along the sides. Apply two more panels to the frame as instructed above, leaving an open panel between each pair. At this point, you will have covered every other panel on the frame. Allow the glue to dry thoroughly.

Step 3

Glue the remaining three vellum cutouts to the frame. You will only be able to clip along the top and bottom wires. Run your finger down the sides of each panel to insure the vellum makes contact with the glue. Once dry, adhere one of the six crinkled paper cutouts to each panel of the shade using double-sided adhesive tape or glue.

4

Step 4

Apply the black trim along the outside bottom of the shade, wrapping it over the wire towards the inside of the shade. You will need to miter the trim at each corner to make it lie flat. Repeat process to apply trim to the top wire. Let dry.

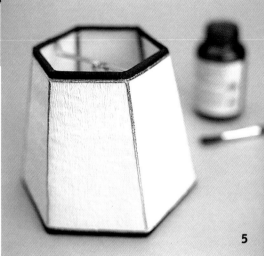

5

Step 5

Glue the silver soutache along the bottom edge of the top trim and to the top edge of the bottom trim. Glue the remaining soutache down the side wires of the frame to cover the seam between the six adjoining panels.

LACED NAUTICAL SHADE

This is a bare minimum shade with clean lines and few fixtures. Yet the shade is as much decorative as it is functional, with brass grommets and hemp lacing contributing to an open, nautical look. Although this shade was designed for a hanging lamp, it can easily be adjusted to sit on a lamp base.

While a variety of papers could be used to make this shade, simple papers work best. Here, a white parchment paper with subtle mottling was chosen, as it mimics the burlap-like fabric of a ship's sail. Just as important is the paper you use to back the shade. This shade hangs without any structural wiring, so the shade must be backed with a sturdy paper to provide support. It should not be so stiff, however, that it cannot comfortably wrap into a loose, drum-like shape.

Materials

- 1 large sheet of lightweight white parchment cut into a 36" x 11" (91 cm x 28 cm) rectangle
- 1 large sheet of heavyweight white parchment or vellum cut into a 36" x 9" (91 cm x 23 cm) rectangle
- Double-stick adhesive tape
- 20 brass grommets
- Inserting grommet handle
- Hammer

- Towel or thick felt
- 5 1/2 feet (1.7 m) of hemp string
- 1 bottom lampshade wire, 12" (30 cm) in diameter
- 13 feet (4 m) of brass wire (28 or 24 gauge)
- 1 brass ring
- Basic craft supplies (see page 146)

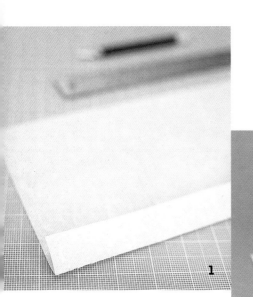

Step 1

On the lightweight parchment, use the back edge (the blunt edge) of a craft knife to score fold lines (top and bottom) parallel to and 1" (3 cm) in from the long edges of the paper. Fold the top and bottom 1" (3 cm) flaps over to the same side of the paper. This will be the front side of the shade.

Step 2

Adhere the stiffer, heavyweight parchment to the backside of the folded parchment, using double-stick adhesive tape. Be sure the folded side faces out. If your heavyweight parchment tends to curve (from being rolled) making it difficult for the paper to adhere, attach paper clips along the edge to hold the papers together. (Once the grommets are in place, it is less important that the two papers adhere.)

Step 3

Use a punch to puncture holes set 3 ¾" (10 cm) apart along the center of the folded bands. Start the first hole 1" (3 cm) in from the far left edge so that you do not end up positioning a grommet where the seam falls.

Step 4

Place a grommet through the front of the shade so that the post sticks out the back. Place the flat, washer-like piece over the grommet post on the backside of the shade. Place an inserting grommet handle into the post and strike with a hammer to seal. Make sure you lay a towel or piece of thick felt between the paper and your work surface before hammering.

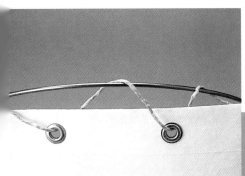

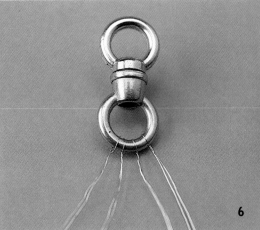

5

6

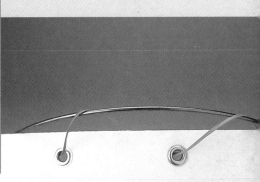

Troubleshooting

When lacing the shade, it can be hard to manage the string and keep the shade hanging an even distance from the frame. One solution is to hold the shade and string in with paper clips so that it won't shift as you lace around the shade.

Step 5

Overlap the opposing short edges of the paper by ½" (.5 cm) and glue to create a cylinder shape. Loop hemp string through the holes and over the wire ring to hang the shade. (See Troubleshooting.)

Step 6

Thread a 34" (86 cm) brass wire through the brass loop, folding the wire in half where it passes through the loop. Bend the wire ends under the lampshade wire and then twist the wire around itself to hold it in place. Use wire cutters or sturdy scissors to clip any extra.

Tip: To adjust this shade so that it sits on a lamp base, use a washer top wire, instead of a bottom wire, from which to hang the shade. The size of the shade is also easily adjusted. Just make sure the length of your paper equals that of the lampshade wire's circumference. Also, check that your grommets are spaced evenly around the shade.

Variation

To add some color to this shade, replace the hemp lacing with sea green gimp.

EXOTIC
PAPER SHADES

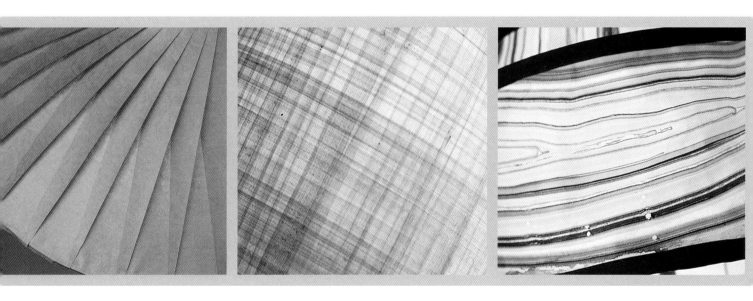

MUSICAL SCRIPT COOLIE

The square coolie is a perfect shape to complement an arts-and-crafts style lamp base. In this project, two papers are used to create an alternating pattern from side to side. Framing the sides with trim accentuates this play in patterns and lends this shade a certain distinction. This particular shade would work well in a study or near a music corner in your house. By simply changing the paper choice, be it in pattern or color, it can go in just about any room, including a bedroom.

Paper Choice

Vellum tends to pucker once the glue adhering it to the frame wires dries. To smooth any rippling or puckering before covering with the paper trapezoids, blow on it with hot air from a hair dryer. The heat will help pull the vellum taut.

Materials

- 1 25" x 38" (64 cm x 97 cm) sheet of medium weight translucent vellum

- 1 19" x 27" (48 cm x 69 cm) sheet of musical print paper

- 1 19" x 27" (48 cm x 69 cm) sheet of antique parchment

- 1 6" x 16" x 11" (15 cm x 41 cm x 28 cm) washer top square coolie frame

- 4 feet (1.2 m) of $\frac{1}{8}$" (.3 cm) wide black soutache trim

- 2 $\frac{1}{2}$ yards (2.2 m) of $\frac{3}{8}$" (1 cm) wide black grosgrain trim

- 4 feet (1.2 m) of $\frac{1}{8}$" (.3 cm) wide black grosgrain trim

- Double-stick Tape

- Basic craft supplies (see page 146)

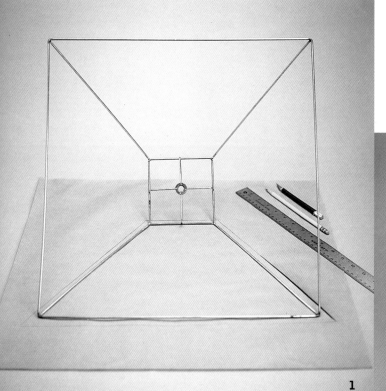

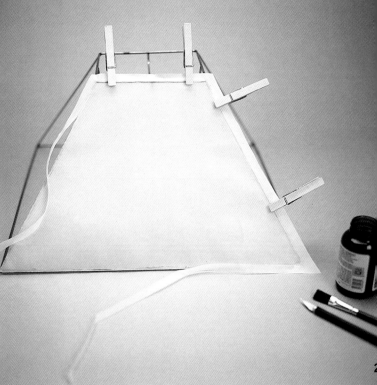

Step 1

Place the lampshade frame over a sheet of scrap paper and draw an outline, allowing for a ½" (1 cm) border. Cut the outline from the scrap paper and use as a template to cut out four vellum trapezoids.

Step 2

Trace a thin line of glue along the wires of one frame side. Spread evenly with your fingertip or a cotton swab. Lay the glue-coated frame side down on top of one of the vellum trapezoids making sure the vellum adheres to the frame smoothly. Trim off excess using a craft knife. You can use clothespins to hold in place while drying. Repeat with the other three sides.

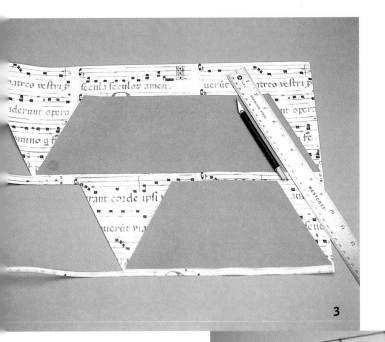

3

Step 3

Place the frame over a scrap piece of paper and draw an outline of one side as in step 1 but without any added border. Cut the outline and then fold in half horizontally. Cut along the fold line so that you have two smaller trapezoid templates—a top and a bottom. Using the top template cut two trapezoids from both the antique parchment and the scripted paper. Do the same with the bottom template. In total, you should have eight trapezoids.

Step 4

Outline the top half of a vellum panel with double-sided tape, matching the outline of the top trapezoid template, as shown. Carefully place one of the top parchment trapezoids over the tape to adhere to the shade. Similarly, outline the bottom half of the panel with double-sided tape. Adhere a bottom trapezoid from the scripted paper to the panel. Repeat around the shade making sure that the papers on the top and bottom panels alternate all the way around.

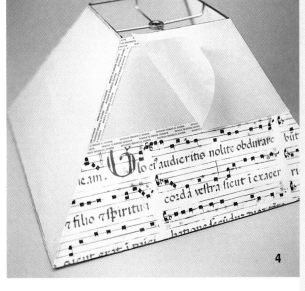

4

On each panel apply a strip of black soutache trim across the center so that it overlaps where the top and bottom trapezoid edges meet. Apply the $^3/_8$" (1 cm) trim in strips over the top and bottom wires of each panel. Run the $^1/_8$" (.3 cm) black trim down the side ribs of each frame, tucking the ends under the top and bottom wires to finish.

STRIPED WHIMSY SHADE

Made with colorful strips of vellum dangling from silver rings, this shade evokes a smart playfulness that conjures up images of carousels, and general whimsy. This is a surprisingly simple shade to make and easily opens itself to variation, be it in the choice of color, pattern, shape of strips, or approach to attaching the strips to the lampshade frame. A clip-on lampshade makes attachment to the base quick and easy. This shade was made for a small base, however any size base can be used. To determine the proportionate size of strips you need for your base of choice, use cardboard or poster board to cut out and hang "test" pieces.

Materials

- eighteen Silver Pierced Earring Hoops, about 1 cm in diameter (or other thin-wired rings that can be threaded)

- eighteen silver eyelets

- one eyelet tool for sealing eyelets

- three 12" x 19" (30 cm x 48 cm) sheets of azure Chromatica translucent vellum (27 lb. weight)

- three 12" x 19" (30 cm x 48 cm) sheets of turquoise Chromatica translucent vellum (27 lb. weight)

- three 12" x 19" (30 cm x 48 cm) sheets of indigo Chromatica translucent vellum (27 lb. weight)

- one 7" (18 cm) clip-on single ring lampshade frame

- Multipurpose spray glue

- Clean, soft cloth

- Basic craft supplies (see page 146)

Step 1

Cover the work surface with paper bags or some other scrap paper that have been cut open so they lay flat to protect the counter from being coated with glue. Place one sheet of azure colored vellum on your work surface and lightly coat with spray glue all over. Take a second sheet of the azure vellum and adhere to the first, beginning in the corner to make an exact alignment. Repeat with a third sheet of the azure. Repeat again with the turquoise vellum and then the indigo vellum.

Step 2

Using the template on page 284, use a craft knife to cut six slats from each of the triple-layered papers so that you have a total of eighteen slats. Using the template as a guide, draw a circle $\frac{1}{2}$" (1 cm) from the top edge of each slat. Cut out the circle on the top of the slat to create a hole.

Troubleshooting

If you cannot find thin-wired earrings that are medium in size, you can always trim larger earrings with wire cutters. Then bend the loop accordingly to tighten the loop.

If your rings do not hold their latch easily, use needlenose pliers to squeeze the backing so that the wire cannot pull out. An alternative would be to apply glue.

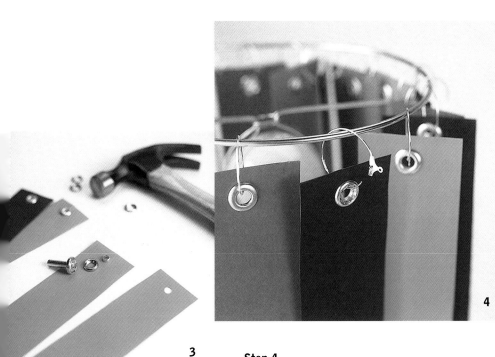

3

4

Step 3

Insert the deep half of the eyelet in the hole so that it is positioned on the front side of the slat. On the backside of the slat, fit the wider, shallow half of the eyelet onto the neck of the deep eyelet half. Place the wide end of the eyelet tool over the shallow half of the eyelet. On a firm, protected surface, use a hammer to strike the tool a few times so that the deep half of the eyelet rolls back over the shallow half.

Step 4

Thread a ring through the eyelet of a slat. Hang the ring on the lampshade frame. Repeat until all the slats are threaded and hanging. Space each ring about 1 ¹/₂" (4 cm) apart on the frame. Place the backing to the ring behind the slat so it is not readily visible.

Variation

To further the element of surprise with this shade, a sprinkling of confetti was trapped between two strips of lilac colored vellum. The strips were then hung from gold lanyard hooks instead of rings. This simple shade can be made with a wide variety of papers. It is important to use a fairly rigid paper that has some weight to it because the slats hang straight and uniformly. Or, if you have a particular paper in mind that is lightweight, use a heavy, stiff vellum or even cardboard for backing. This capricious style of shade particularly lends itself to vibrant colors.

PINK BOA SHADE

If only Barbie could wear this! It's hard to get any more playful than dressing a gathered pink paper shade with feather boa trim. While this diaphanous shade might find itself perfectly in place on a little girl's dresser, it might also find a perfect spot on a vanity or on display in the home of anyone with a flair for a little tongue-in-cheek kitsch. The machine-made paper used to create this shade is composed of hair-like rayon threads with gold and silver foil flecks dispersed randomly throughout. It is extremely lightweight and semi-transparent (more so when light is shined from behind it). Assembly of this shade requires elementary sewing skills. If you have never hand-basted before, you can learn here without worry. It gets no harder than being able to thread a needle.

Paper Choice

As if made from fabric, this shade gathers softly and loosely, so it's important to select a paper that is as supple and resilient as fabric. Rayon-based paper works ideally. There are also a number of soft lace papers that would work well for this shade. The paper you choose should be so soft that you could crumple it in your hand without its creasing or holding the shape when you release it.

Materials

- 1 sheet of 25" x 37" (64 cm x 94 cm) medium-weight translucent vellum

- 2 sheets of 21"x 31" (53 cm x 79 cm) burgundy lace rayon paper flecked with gold and silver

- 2 yards (1.9 m) of white feather boa, about 2 inches (5 cm) thick

- 1 round top washer lampshade wire 3 ½-inch (9 cm) diameter

- 1 round bottom lampshade wire 10-inch (84 cm) in diameter

- 1 needlepoint needle

- 1 sewing needle

- White sewing thread

- Basic craft supplies (see page 146)

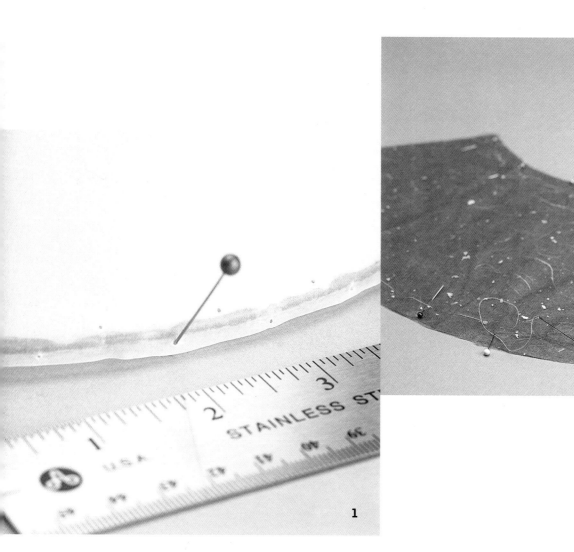

2

1

Step 2

Cut out both sheets of rayon paper using the template on page 287. Pin the two sheets together on top of one another. Thread a long strand of white thread through the sewing needle. Do not make a knot at the end of the thread. Baste stitch along the inside curved edge of the pink paper leaving at least six inches (15 cm) of string (not knotted) hanging from each end. The stitches should be about $3/4$-inch long (2 cm), and about $3/4$ inch (2 cm) in from the paper edge. Repeat basting along the outside curved edge.

Step 1

Using the template on page 286, make a cone shaped shade using the vellum paper. The top wire is meant to fit inside the vellum cone with the paper hanging over about $1/2$ inch. Likewise, the bottom wire will sit slightly inside the cone. Use a needlepoint needle to puncture two rows of holes along both the top and bottom edges of the cone to make stitch guides for step 4. Along the top edge, the puncture holes should be staggered about $1/2$ inch (1 cm) apart. Along the bottom edge, the holes should be staggered about one inch apart.

4

3

Step 3

To create gathers, hold onto one end of the thread used for basting the inside curve and push the paper towards the center edge of the curve. Repeat on the other side to complete the top gather. Then repeat on the outside curved edge to make the shades bottom gather. Use clothespins to wrap the gathered paper around the vellum shade, aligning seams accordingly. Adjust gathers as needed so that they balloon around the shade evenly. The seams of the gathered paper should overlap at least $\frac{1}{2}$ inch (1 cm) and should hang loosely. Run a line of glue along the top and bottom edges of the vellum lampshade to adhere the gathered paper shade cover. Glue in between the clothespins. Once dry, remove the clips and finish gluing. Let dry.

Step 4

Clip away the excess basting thread. Beginning at the back seam of the shade, sew the feather boa to the top edge, using the punch holes as guides for zigzag stitching. Once the boa has completely wrapped around the top edge, clip it. Use the remaining boa to stitch to the bottom edge of the shade.

CRUSHED LEATHER LOUVRED SHADE

The simple technique of overlapping strips of paper puts a modern twist on the pleated shade (and absolves the crafter from the meticulous challenge involved with pleating). Because of the wide-brimmed shape of the shade, the strips of paper splay out at the bottom. This creates a pattern of shadows that tapers to a point at the shade base, like warm rays of sun bathing the lamp.

This shade is dramatically wide-brimmed and can take on a variety of base types, depending on the effect you wish to create. Here a tall, narrow base was used to create a neckiness to the ensemble. A stout, round ceramic base would work equally well.

The paper used here is manufactured to resemble crushed leather. Like so many artistic papers, this one takes on a whole new effect when placed in front of light. In this case, the texture of the paper—like that of soft, crushed leather—becomes accentuated.

This shade is composed of multiple, narrow slats of paper, so it is delicate. Be sure to place it where it is least likely to get bumped or snagged.

Paper Choice

There are a lot of lines and angles to this shade, so a simple, plain paper works best. When selecting the paper, hold it up to the light. It should be slightly translucent, ensuring that the shadows where the slats overlap will be pronounced when the lamp is lit.

Materials

- 2 sheets of 19" x 27" (48 cm x 53 cm) heavyweight translucent vellum

- 2 sheets of 21"x 31" (53 cm x 79 cm) amber "faux crushed leather" paper

- 1 washer top lampshade frame wire, 4" (10 cm) in diameter

- 1 bottom lampshade frame wire, 18" (46 cm) in diameter

- Basic craft supplies (see page 146)

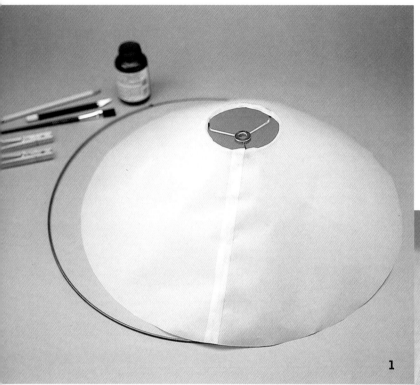

1

Step 1

Use the template on page 290 to cut the vellum sheets into two arcs. Glue one of the plain vellum arcs to the frame wires making a cone shape. Trim any paper that hangs beyond the wires. In accordance with the pattern of dotted lines on the template, lightly pencil in guidelines on the remaining unattached vellum arc. Lay the arc-penciled side facing up-on your work surface on top of a large sheet of scrap paper.

Step 2

Use a craft knife to cut the crushed leather paper into thirty-three slats, each 2" x 9 ½" (5 cm x 24 cm). At each end of the paper slats, mark a "v" shape at the halfway point.

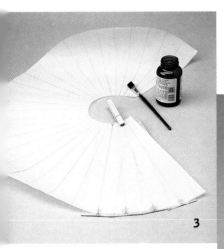

3

Troubleshooting

If the paper you choose is particularly thin or lightweight, you may wish to reinforce the paper slats. To do so, simply cut thirty-three slats, each 4" x 9 ½" (10 cm x 24 cm). Fold them in half lengthwise and position on the shade so that the folded edge is what sits on the outside.

Step 3

At one far end of the vellum arc, slide a slat ("v" side up) under the arc and align the mid-points of the slat with the ends of the first penciled line on the arc. The slat should project slightly beyond the curved edges of the arc. Spread a small dab of glue or piece of double-stick tape near the center base of the slat to tack in place to the arc. Leave the top end unglued. Repeat, aligning and gluing all but two slats to the arc up to the last two lines. Do not glue the top ends of the slats in place.

4

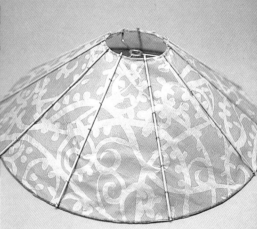

Step 4

Carefully flip the slatted arc over and wrap it around the shade. Run a thin line of glue three-quarters of the way around the top edge of the vellum shade to adhere the top of the slatted arc to the shade. Do not run glue under the top edge of the first five slats. They need to remain open for the missing two slats to be inserted. Hold the top edge of the shade in place with clothespins if necessary while the glue dries. Insert the remaining two slats into position, the last of which will overlap more than average. Glue the two remaining slats and the first five in place at the top edge. Dot glue beneath the bottom corners of the slats to complete.

Variation

You can vary the look of this shade by making a solid arc with a patterned paper of choice (that is, skip the steps of covering the shade with overlapping slats). Instead, stitch wooden skewers to the shade to create an umbrella-like feel.

CORK LAMPSHADE
with Faux Leather Stitching

Pressed cork paper is one of the most fascinating papers available—if just for the fact that it doesn't seem like paper at all! It has the cushioned feel of cork, the speckled markings of cork, and the warm wood-like face of cork. And this paper works great for shades. It is sturdy and decisive in its presence. However, pressed cork paper is not translucent, so do not choose this shade for a space where you really need light to splay out.

The available size of pressed cork paper—20" x 30" (51 cm x 76 cm)—dictates that two pieces are needed to wrap around the entire shade. Because of the variegated patterning of the cork, however, the double seams will go unnoticed.

Materials

- 1 strip of 9" x 33" (23 cm x 84 cm) craft paper
- 1 sheet of 20"x 30" (51 cm x 76 cm) pressed cork (cut into two 9" x 17" (23 cm x 43 cm) rectangles)
- 1 top lampshade wire, 10" (25 cm) (either clip-on or washer style) in diameter
- 1 bottom lampshade wire, 10" (25 cm) in diameter
- 4 yards (3.5 m) of brown leather-like cord
- Basic craft supplies (see page 146)

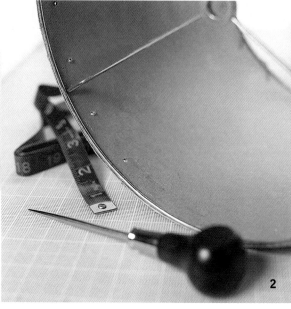

Step 1

Using the craft paper, follow step one on page 174 to make a drum-shaped shade 9" (23 cm) in. Run a line of glue almost halfway around the top edge of the drum shade. Adhere one sheet of the cork paper to the shade, leaving about an inch along one edge free (as shown). Repeat along half of the bottom edge-minus 1" (3 cm). Let dry. Run a line of glue around the remaining top edge of the drum shade. Adhere the second cork sheet, tucking one of its edges under the 1" (3 cm) long free edge of the first cork strip. This way, one end of each strip overlaps the other. Glue the bottom edge as well. Run glue or double-stick tape under the side seams to seal.

Step 2

Using an awl or other sharp pointed object, such as a needlepoint needle, puncture holes around the top and bottom edge of the shade. The holes should be positioned $5/8$" (1.5 cm) in from the wire edges and $1\,5/8$" (4 cm) apart. The easiest way to measure around the shade is by using a flexible seamstress tape.

Step 3

Thread the leather-like cord through a hole coming from the front of the shade through to the back (leave enough slack to tie off ends). Pull the cord over the wire edge and thread again through the hole to the immediate left, again coming from the front. Repeat around the shade. You may find that the tip of the cord occasionally needs to be snipped at an angle with scissors to be able to thread the cord through with ease.

Variations

In addition to traditional ribbon trim, there are boundless options for adding a decorative finish to your paper shade. The following are just a handful of ideas using actual paper trims or common paper embellishments:

Raffia is a perfect partner to so many handmade papers. Here 6 strands are simply twisted together to trim along the edge of a mulberry paper.

Alternatively, run a single strand along the edge, looping the raffia at intervals for a soft, decorative touch.

4

Step 4

Once you have stitched around the entire shade, tie the string ends together in a knot on the inside of the shade.

5

Step 5

For a subtle decorative finish, between the first and last stitches overlap two small stitches to make an "X" shape to complete.

Troubleshooting

• *When working with pressed cork paper, it is important that the grain runs vertically on the shade. If you bend the cork so that the grain wraps around the shade, the cork will split. If you are uncertain how the grain runs on your paper, try rolling a test swatch. It will naturally roll parallel with the grain. It will resist and the cork will crackle if rolled perpendicular to the grain.*

• *Use the backside of the paper, which is white, to mark measurements. Pencil marks easily get lost in the maze of texture and speckling on the cork side.*

Tiny paper flowers are gathered side-by-side to make an ornate, three-dimensional trim. An awl was used to puncture holes along the paper edge to thread the wire stems of these flowers through to the back side of the paper.

For a more organic look, fossilized citrus leaves were overlapped and held down with a coordinating bark-colored paper ribbon.

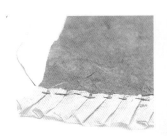

Soft unryu paper in peach was folded to create a delicate pleated trim with orange raffia stitching as an accent.

A strip of Italian spiral-patterned paper adds interest to an otherwise plain paper backdrop. Cut paper into a narrow strip and fold in half with diamond cutouts falling at intervals along the fold.

DOUBLE-DECKER MARBLED LAMPSHADE

Here two shades, topped one on the other, make one shade—and a surprisingly different one at that. To add to the fun, the shades are made with Thai hand-marbled paper that makes this shade all the more exotic.

There is nothing tricky about crafting this shade. You simply make a basic drum-shaped shade, albeit much shorter and wider than an average shade, in order to create a more dramatic effect. Then you prop a smaller shade on top. No unusual hardware is required to suspend two shades over one base. You simply need to line the top shade with a heavyweight vellum or parchment so that it can stand up on its own on top of the spokes of the bottom shade.

One distinctive touch with this shade is that two layers of trim are applied. The first is applied in a traditional manner to wrap over the shade and its structural wires. The second is laid flat around the shade edge so that it creates a thick band with a crisp edge along the top and bottom edges of the shade.

This shade will work equally well with less dramatic papers. You can use the same paper for each shade level or try mixing and matching patterns or varying colors. Just about anything goes.

Materials

- 1 large sheet 25"x36" (64 cm x 91 cm) of heavyweight translucent parchment or vellum for shade lining

- 1 large sheet 25"x36" (64 cm x 91 cm) of Thai marbled paper

- 16" (41 cm) washer top lampshade wire

- 16" (41 cm) bottom lampshade wire

- 12" (30 cm) washer top lampshade wire

- 12" (30 cm) bottom lampshade wire

- 10 $\frac{1}{2}$ yards (9.6 m) of black $\frac{5}{8}$" (1.5 cm) wide grosgrain trim

- Basic craft supplies (see page 146)

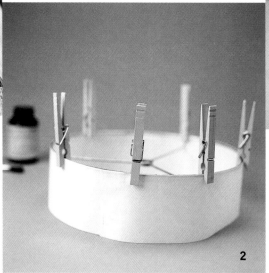

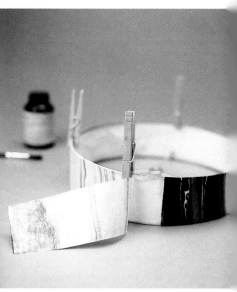

Step 1

Using a craft knife, cut the paper for the shade lining into two rectangles: one measuring 8" x 52" (20 cm x 132 cm) and the other 3" x 39" (8 cm x 99 cm). Do the same with the marbled paper. It is unlikely you will find either paper in 52" (132 cm) lengths. To remedy, make the large rectangle with two pieces measuring 8" x 26 ¼" (20 cm x 26.5 cm). Adhere with glue or double-stick tape along one short edge by over-lapping ½" (1 cm). For the marbled paper, try to overlap with pieces that have similar edges so as not to disrupt the pattern too dramatically, as shown here.

Step 2

Glue a long edge of the 8" x 52" (20 cm x 132 cm) sheet of to the 16" (41 cm) top wire. Clip with clothespins while the glue dries. Repeat the process with the 3" x 39" (8 cm x 99 cm) lining and the 12" (30 cm) top wire. Once the glue has dried, glue the bottom edges of each lining to the corresponding bottom wires.

Step 3

Cover large drum and small drums with marbled paper using double-stick adhesive tape or glue. If gluing, hold in place with clothespins until dry.

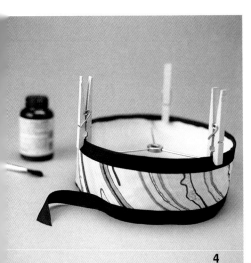

4

- *This project uses quite a lot of trim. If you're shy on trim, cut a supple paper into ⁵/₈" (1.5 cm) wide long strips and apply as the under layer of trim that wraps inside and over the top and bottom frame wires.*

- *When applying the second layer of trim to the shade, set the shade at eye level on the lamp to achieve accuracy. Hold in place with the finial and gently spin the shade as you work around it.*

Step 4

Glue trim along the top and bottom edges of each drum, folding over the edges to cover frame wires inside the shade. Hold in place to dry with clothespins. Once dry, apply a second layer of trim around the outside top and bottom edges of both shades so that the trim edges fall flush with, rather than fold over, the edges of the shades. Use double-stick tape or glue. If using glue, hold trim in place with clothespins to dry.

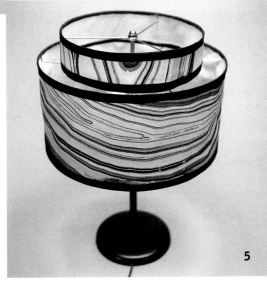

5

Step 5

Secure the large drum to the lamp base, securing in place with a finial. Set the small drum on top, propping it on the spokes of the top wire of the large drum.

Variation

If a double-decker shade is not daring enough, why not push the limit and make a triple-decker? For this shade, the third tier was made with 8" (20 cm) diameter frame wires.

VOTIVES AND NIGHT-LIGHTS

DRAGONFLY NIGHT-LIGHT

A night-light is one of the few lights in a home that almost always shines alone and in the pitch dark of the night. So why not make it a light that is pleasing to the eye—especially to the weary eye.

This particular shade is shaped like a sconce to curve well around the night-light bulb. It is also large enough to cover a standard outlet frame. For this project, be sure to pair the shade with a night-light fixture that takes a cold-burning bulb. You can find this kind of night-light in a hardware or general housewares store.

Night-lights are particularly conducive to construction with sculpted wire because they are small and do not need a heavy, rigid wire for support. You might wish to buy extra wire, however. These shades can be so inspiring to sculpt with that you might just end up making more than one.

Materials

- 1 sheet of 9 $\frac{1}{2}$" x 11" (24 cm x 28 cm) vellum with gold dragonfly pattern
- 18-gauge pearlized gold wire cut into the following lengths: 26" (66 cm), 9" (23 cm), and 7 $\frac{1}{2}$" (19 cm)
- 24-gauge pearlized wire cut into the following length: 18" (46 cm)
- 1-inch (3 cm) wide gold sparkle abaca ribbon or other trim, cut into the following lengths: 9" cm) and 7" (18 cm)

- Standard and needle-nose pliers
- Wire cutters
- 28 oz. (.8 liters) aluminum can or 4" (10 cm) diameter aluminum can
- 3 $\frac{1}{2}$" (9 cm) diameter can
- 1 night-light with a cold-burning bulb
- Double-sided tape
- 1 gold, finish nail
- Basic craft supplies (see page 146)

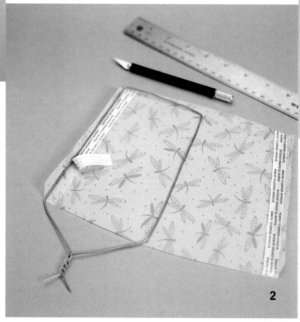

Step 1

Using a square edge such as the corner of a table, bend the 26" (66 cm) wire to conform to the template shape on page 288. With the pliers, grip the wires just below the point at which they intersect, making an "x." Hold the wire tips together with the needle-nose pliers and twist until they are wrapped around one another.

Step 2

Cut the shade cover from the vellum paper using the template on page 289. Using the back-side of a craft knife, score the backside of the paper $\frac{1}{2}$" (3 cm) in from the two straight sides. Place a strip of the double-sided adhesive tape along the inside edges of the scored lines. Position the wire frame so that the long straight edges of the frame sit on the paper's scored lines. Fold the paper over the wires to adhere to the tape and lock in the wires.

Step 3

Bend the 18-gauge 9" (23 cm) wire halfway around the 28 oz. (.8 liters) can to create a curve. Insert the curved wire inside the shade so that it sits $\frac{3}{8}$" (1 cm) below the top curved edge. Glue to the inside of the shade using clothes pins to hold the wire in place as the glue dries. Using the needle-nose pliers, wrap the wire tips around to the front sides of the shade. Wrap the 7 $\frac{1}{2}$" (19 cm) wire halfway around the smaller aluminum can. Adhere to the bottom curved edge of the shade using the same method as for the top wire.

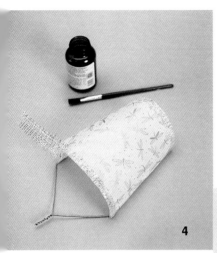

Step 4

Using glue, adhere the 7" (18 cm) gold trim to the bottom rounded edge of the shade so that ¹/₂" (3 cm) of the trim hangs over the outside edge. Repeat with the 9" (23 cm) trim on the top rounded edge.

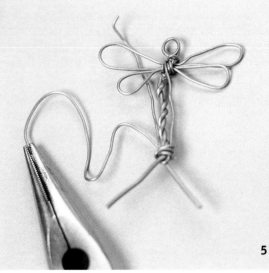

Step 5

Sculpt a design of choice with the 18-gauge wire and attach it by lassoing it to the frames twisted top wires. Position the finished shade over the plugged-in night-light, using a finish nail to hold the shade in place.

Troubleshooting

While you can find 18-gauge and 24-gauge wire in hardware stores, they are not necessarily designed for crafting. It can be particularly difficult to smooth out any kinks. The same gauge wire found in craft stores is designed for sculpting, so it is much more malleable. To get clean, round shapes, use a pencil or other round object around which to wrap the wire. To smooth out kinks, firmly press the wire against a pen or other round object with your thumb and pull the wire through. Repeat if necessary.

Variation

For this project, you can sculpt just about any design you would like to lash to the nightlight's wire frame. Perhaps you would like to coordinate it with wallpaper or a room theme, such as leaves. Just make sure the wire-sculpted design includes a centrally located loop to hang the shade by.

RIPPLED FOIL VOTIVE

The natural beauty of unique papers can be used not only to diffuse light but also to reflect it. Foil papers prove ideal for this effect and come in a variety of metal tones and even textures.

A hammered copper paper was used in this project for its unique textural component and the added warmth copper contributes to the candlelight. A dark paper with copper marbling was selected for backing to give the votive added structure and a midnight feel.

The basic shape of the votive is similar to that of an electric drip coffee filter and is very simply made by gently molding the paper around a round, tapered object. The grass wreath surrounding the base finishes the presentation, while displaying your work on a candle stand pronounces the delivery.

Materials

- 1 sheet of embossed foil paper cut into a 9" x 9" (23 cm x 23 cm)

- 1 sheet of black construction paper with metallic marbling or other sturdy weight paper cut into a 9" x 9" (23 cm x 23 cm)

- Small bunch of dried wild grass (available at craft stores)

- 9 strands of 5" (13 cm) copper or natural raffia

- 1 medium size permanent marker (or object of similar shape and size)

- 1 square candle stand, 5"x5" (13 cm x 13 cm)

- 1 drip-proof candle, 3" tall x 2" wide (8 cm tall x 5 cm wide)

- 2 ½" (6 cm) round glass candle plate

Step 1

Glue the backsides of the black marbled paper and foil paper together. Cut out an 8 ½" (22 cm) diameter circle from the glued papers. Using a pencil, lightly draw a 2 ½" (6 cm) circle in the center of the paper circle on the foil side. Place the marker tip (cap on) at the edge of the inner circle and gently mold outer circle around the marker. At the edge of the inner circle the paper should just barely mold around the marker tip; do not crease it. The curves should be soft and less pronounced towards the center of the disc.

Step 2

Place the marker on the backside (black paper side) of the disc about 1" (3 cm) to the right of where it was previously positioned. As in step 1, mold the paper around the marker to create another, opposing, curve. Repeat around the disc to create a rippled effect.

4

Troubleshooting

Foil papers are not so forgiving when you smudge even the least bit of glue on them. To minimize the risk of getting any glue on the paper, place the foil face down on a clean surface and glue the black paper on top of it (instead of vice versa). If there is any glue on your hands, it will smudge on the black paper, where it is less likely to show. To be safe, always keep a damp cloth nearby to keep your hands clean while applying glue and handling papers.

Step 4

Once you've made a 16" (41 cm) chain, wrap it into a circular shape, lashing the top 1" (3 cm) of the first shock of grass with the base of the last to form a wreath with a 3 ½" (9 cm) diameter. Trim off ends of the raffia ties and center the wreath on the candle stand. Position the rippled foil "basket" inside the wreath and place the glass candle plate inside the votive. Set the candle on top of the plate.

Variation

In addition to foil papers, there are a number of other papers to choose from that have reflective properties, such as this "opalescent" paper. Gold leaf is another excellent alternative if you are adept at working with it.

3

Step 3

Make a chain of dried grass. To make the chain, overlap one shock of dried grass about 1" (3 cm) below the tip of the first. Using a strand of raffia, tie the shocks of grass together at one-inch intervals. Trim off any woody stems as you go along.

STARLIGHT HANGING LAMPSHADE

Perhaps the most ethereal and yet tactile of papers are Japanese lace papers. Their intricate webs of threading suggest a delicacy unlike any other. Yet these papers can be crumpled up and tugged upon and still maintain structural integrity. The "see-through" nature of lace papers works particularly well with this hanging shade, in which miniature lights twinkle through the paper's open weave. Consider this more a decorative "night-light" than a functional source of illumination.

Lace papers do not adhere well with glue to shade wires, particularly when there is no backing to the shade, as is the case for this shade. For this reason, the paper in this project is treated like a fabric and actually stitched to the wires.

Materials

- 1 washer top lampshade wire, 10" (25 cm) in diameter
- 1 bottom lampshade wire, 10" (25 cm) in diameter
- 1 clip shade adapter
- Twist ties
- 1 strand of small white Christmas tree lights
- 1 large sheet of steel blue Japanese lace paper, cut into a 32" x 20" (81 cm x 51 cm) rectangle
- Raffia
- Double-stick tape
- Sewing Needle
- Steel-blue thread

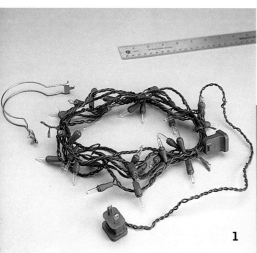

1

Step 1

Wrap the strand of lights into 8" (20 cm) length loops. Use a twist tie to pull together the strands at one end of the loop. At the other end, sling a clip shade adapter underneath.

2

Step 2

Use double-stick tape to adhere the first 10" (25 cm) of the 20" (51 cm) sides of the rectangle to form a cylinder shape. Leave the remaining 10" (25 cm) unattached for the time being.

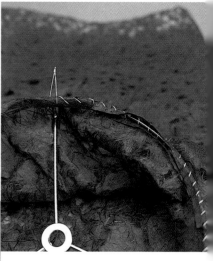

Step 3

Fold in by ½" (1 cm) the top edge of the cylinder. Fold again by ¼" (.5 cm) over the top frame wire and whipstitch to attach, here shown in a contrasting thread to better exemplify. Also, whipstitch the bottom lampshade wire to the paper cylinder so that it sits 10" (25 cm) below the top wire.

Tip: To make sure the bottom wire falls evenly 10" (25 cm) below the top wire, you may find yourself constantly keeping measure. One shortcut is to use a fabric pencil to mark the paper 10 ¾" (27 cm) below the top edge before forming the cylinder in step 2. (The additional ¾" [2 cm] takes into account the folding in step 3.)

Troubleshooting

*This shade has little or no rigidity.
So construct as much of the shade as
possible before inserting the strand of
lights; otherwise, you will encounter
some awkward handling. Finally,
make sure your strand of lights is
the type that does not heat up. As
with any shade, keep the paper out
of direct contact with any bulbs.*

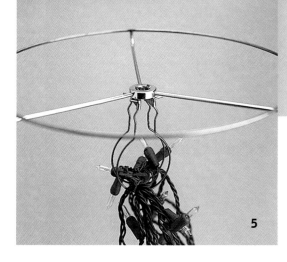

Step 4

Tie strands of raffia to the three spokes in the top wire and knot together at the ends to form a loop from which you can hang the shade. Glue the raffia around the shade where the top and bottom wires are positioned.

Step 5

Attach the strand of lights inside the shade by clipping the adapter to the washer top wire, seen here with the shade removed to better illustrate how the adapter attaches.

Step 6

Pull the loose bottom 10" (25 cm) of the shade together and wrap numerous times with raffia to gather, tying to hold in place.

Variation

As seen here, you can make this shade with an open bottom. It can be used as a hanging shade or as one set on top of a lamp base. The bottom edges of the shade shown here were clipped to create a scalloped effect.

CUTOUT TEA LIGHTS

There is elegance in simplicity. These basic tea lights embody that. Here, a cutout of a stylized tree in a straight lined design provides the only decorative element. With cutout patterns, straightforward design is often the best. Create your own or refer to the templates on page 285. As with any paper votive, it is imperative for safety's sake that the candle is encased in a glass holder.

Materials

- 11" x 14" (28 cm x 36 cm) card stock or other heavy-weight paper
- 1 small piece of white vellum
- Double-stick tape
- Craft knife

Step 1

Determine the height and width you want the shade to be. Here, the shade is 8 1/2" (22 cm) tall and 3" (8 cm) wide. To calculate the size, measure the diameter of the glass candleholder and add at least 1/2" (1 cm) to get the width of each side of the shade. (This allows the minimum 1/4" (.5 cm) breathing room on all sides of the holder.) Multiply the width times four; then add an additional 1/4" (.5 cm) to create a flap for sealing the shade. For the shade here, an 8 1/2" x 12 1/4" (22 cm x 12.5 cm) rectangle was cut from the card stock.

Step 2

With a craft knife, score the rectangle into four equal panels leaving 1/4" (.5 cm) strip at the end. For this project, the paper was scored to make four panels, 8 1/2" x 3" (22 cm x 8 cm) with the fifth panel measuring 8 1/2" x 1/4" (22 cm x .5

cm) at a far end). Then, trace the pattern on page 285 on each panel and cut along the lines.

Step 3

Back each cutout with a piece of vellum using double-stick tape or a glue stick. On the front, glue a 1/4" (.5 cm) border of a different colored paper around the design to "frame" it. Fold along the scored lines, and adhere the 1/4" (.5 cm) flap to the inside of the shade to complete.

Variation

For a quick set of tea lights, cut a square out of each panel to frame a piece of printed, translucent paper such as the sea shell vellum seen here (above, right).

CUT AND PIERCED PARTY SHADES

The flicker of candlelight dancing through cutout and pierced designs on a paper votive will add party enchantment as the centerpiece to a dinner table or as a background accent in a room. To pierce paper, use an awl (a craft tool specifically used for piercing) or improvise with needles sewing, darning, knitting—or even some nails. The paper should have enough structure to permit a clean puncture. Black, fibrous papers with more rigid paper to allow a clean piercing. Setting the paper on a surface that yields to the point of the awl or other sharp object, such as an old towel, a blanket, or, with a small project, a few layers of thick paper towel.

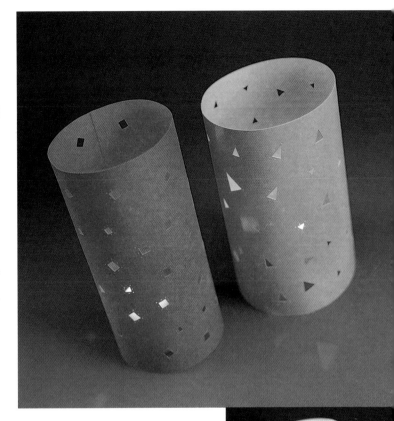

Materials

- 1 sheet of heavyweight red or orange paper (see step 1 for specifics on dimensions)

- Double-sided adhesive tape

- Glass encased votive candle

- Awl (see above for alternative tools)

- Old blanket or towel

- Basic craft supplies (see page 146)

Step 1
From the red or orange paper, cut out a rectangle that measures the same height as the candle-holder and 2" (5 cm) longer than the candle-holder's circumference.

Step 2
Using a craft knife, cut out ¹/₄" to ¹/₂" (.5 cm to 1 cm) squares or triangles from the paper. Do this freehand or make a cardboard template.

Step 3
With an awl or similar pointed tool, pierce holes at each corner of the squares and triangles. Also pierce a hole centered along each side.

Step 4
Finally, use a strip of ¹/₂" (1 cm) wide double-stick tape to close the rectangle into a cylindrical shape; overlap each edge exactly ¹/₂" (1 cm) for a complete seal and a neat finish.

Variation
Use hand punches to create a design like the one seen here, which uses ¹/₈" (.3 cm) and ¹/₄ inch (.5 cm) circles. Since hand punches will only reach about an inch or two (3 cm to 5 cm), ³/₄" wide (2 cm) strips of paper were punched and layered on vellum (above, right).

FOLDED STAR PAPER VOTIVE

Here, celestial, spiral lace paper is girded with wire mesh to fold into a star-shaped candle votive. The trickiest element to this simple shade project is working with the wire mesh. It's prickly, so wear protective gloves. It also can be brittle and easily broken, so you cannot make sharp creases when folding it into the star pattern. To avoid these problems, cut out the paper-covered wire mesh slightly longer than might be needed to allow room for error. Simply trim the excess with wire cutters or old shears.

Materials

- Dark blue paper
- Spiral lace paper
- Fine wire craft mesh
- Spray adhesive
- Work gloves
- Wire cutter or old scissors

Step 1

Cut a 7 ½" x 24" (19 cm x 61 cm) rectangle of wire mesh, using a wire cutter or old scissors. Wear work gloves.

Step 2

Cut a piece of spiral lace paper the same size as the mesh. Cut two strips of blue paper, 1" x 24" (3 cm x 61 cm), for the top and bottom border. Using spray adhesive, affix the borders to the top and bottom of the spiral paper. Next, affix the paper to the mesh with the spray adhesive. Be sure to protect your work surface with newspaper or scrap paper.

Step 3

Fold ½" (1 cm) of the top and bottom edges of the paper-coated wire to the back (see headnote). Fold one of the side edges ¼" (.5 cm) to the back. Measure 2" (5 cm) in from the crease of the folded side edge and fold the mesh in the opposite direction. Continue folding, alternating the direction each time for an accordion fold. Gently fold ten times. The tenth fold should be only ½" (1 cm) wide, so you may need to trim any excess. Glue this flap to the inside of the opposite edge to create and close the star.

Variation

Use a solid piece of color-blended translucent paper for an instant design, like the "sunrise" unryu rice paper used here. To make a triangular shape, fold the mesh three times, with the last fold being ½" (1 cm). Close the triangle as described in Step 3 above.

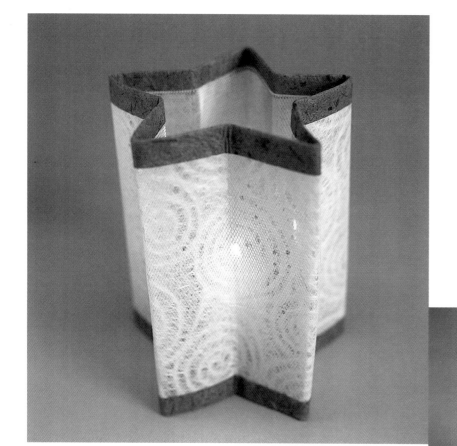

TISSUE PAPER HURRICANE

If you don't have an old glass hurricane shade hanging around in your basement or storage closet, you can certainly find one second hand. Making this shade does not require skill with a craft knife. All you have to do is tear the paper into pieces. The torn edges blend together visually for a soft look. And a third color is naturally introduced when pieces of paper made from two different colors overlap, as seen here. Because it is very difficult to handle wet tissue paper, be sure to apply glue to the hurricane shade, not to the paper. In the variation, since the tissue, when dry, is very translucent, the foil glitter still sparkles even when covered.

Materials

- 1 glass hurricane shade
- Collage glue (clear and fast drying)
- 2 sheets of tissue paper in two or more colors (or any other sheer, transparent papers)
- Foam brush

Step 1
Tear tissue paper into large and medium-sized pieces.

Step 2
Apply glue with a foam brush to a small area of the hurricane shade.

Step 3
Apply pieces of the tissue over the glue, and gently smooth out with your fingers. Continue applying glue and adding tissue to cover the vase, overlapping the edges to blend the colors.

Variation
Completely cover a shade with very large pieces of blue tissue and let dry. Buy confetti, or make some using decorative hand punches. Gold and silver foil paper punched with crescent, star, and circle hand punches were used for the luminary seen here. Apply a thin layer of glue to a small area of the shade and sprinkle with confetti. Continue around the shade, working in small sections. Use a toothpick to adjust any of the confetti. Apply another layer of large pieces of tissue over the confetti (above, right).

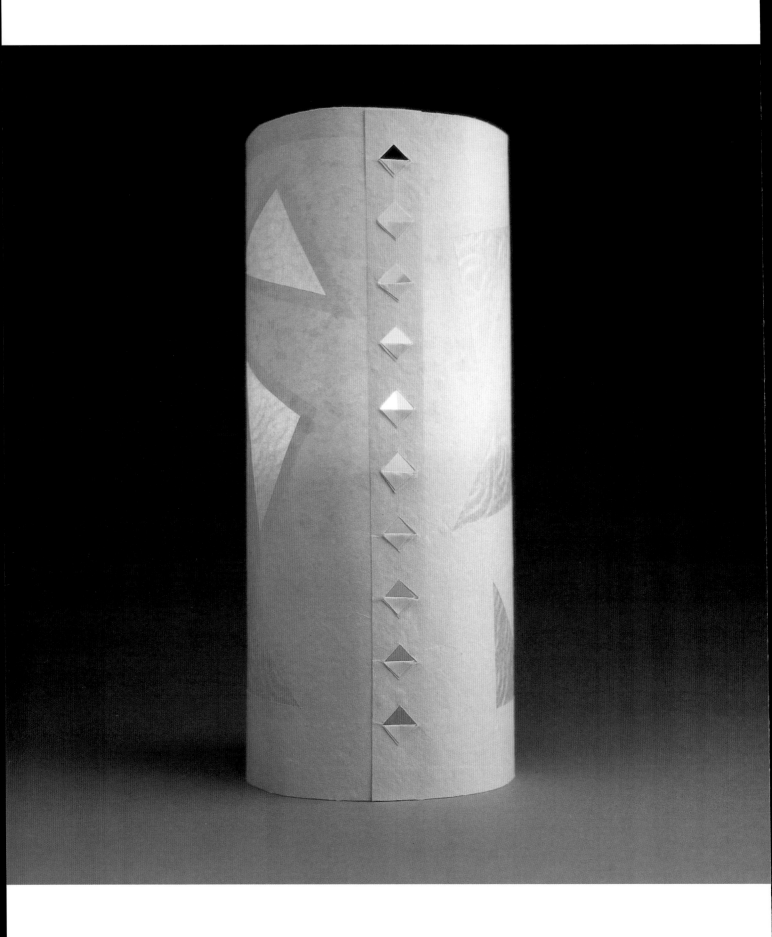

- 12" x 16" (30 cm x 41 cm) sturdy translucent paper for shade

- small sheet of translucent paper for triangle accents

- stiff paper to use as a pattern

- scissors

- craft knife

- cutting mat

- PVA glue

- cotton swab or narrow paintbrush

- bone folder

- pillar candle

JAPANESE GLOW LANTERNS

Simple geometric shapes add a little play to these easy-to-make paper lamps. Choose two papers that will diffuse the light differently. That subtle contrast will make your lamp more interesting. Use a sturdy medium-weight paper for the shade, one that has enough body to stand firmly on its own. For the triangle accents, you can use a lighter weight paper since the firmer base paper will support it. The construction is simple—the paper is merely notched along the overlapping edges and folded over. The notches hold the cylinder together firmly while adding a decorative cut that releases pure light.

STEP 1 Prepare your papers. Cut a 12" x 16" (30 cm x 41 cm) piece of sturdy paper; this will be your shade. Cut several triangles from the accent paper. It's best to make a pattern and use that to keep your triangles uniform in size. Once the triangles are cut, trim 3/8" (9.5 cm) off each edge of your pattern, and use the remaining smaller triangle to mark and cut triangles from your shade paper.

STEP 2 With the front of your shade paper facing down, fold a 1/2" (1 cm) hem in the top and bottom long edges. Next, use a cotton swab or a narrow paintbrush to spread a thin line of glue around one of the cut triangles on the shade. Lay a paper triangle, front side down, behind the cutout. Press the edges of the paper together to adhere. Repeat to cover each of the cut holes with a paper triangle.

To further enhance a crumpled paper, immerse the ball into a diluted tea bath. The softened fibers at the folds will absorb the dye more easily than the rest of the sheet and enhance the crumpled effect.

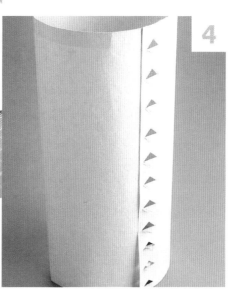

JAPANESE GLOW LANTERNS

VARIATION

Before you start your project, crumple the paper for your glow lamp to give it extra texture. The folds will weaken the paper, allowing more light to shine through those areas. Fold and unfold the paper gently so it doesn't tear, and allow the paper to flatten again after unfolding by weighting it under heavy books or by ironing it using a pressing cloth.

STEP 3 With the paper still lying flat, lift the left (short) edge and fold it over to align with the right edge. Using care not to crush the curved side of the paper, carefully mark and cut five 1/2" (1 cm) upside down V shapes through both layers of paper and evenly spaced along the short side of the papers.

STEP 4 Open the paper and reroll into a cylinder shape with the upside down V shapes aligned again. Push the cut V from the inside of the shade, folding out and down so that the papers interlock. Glue the V tabs in place to secure your glow lamp's cylindrical shape.

- two 8 ¹/₂" x 11"
  (22 cm x 28 cm) sheets
  of turquoise and olive
  green vellum

- flower-shaped paper punch

- sturdy cardboard or mat
  board

- scissors

- white acrylic paint

- craft knife

- metal ruler

- pencil

- PVA glue

PIN-PRICKED VELLUM
NIGHT-LIGHT SHADES

This glowing night-light is made from punched and layered vellum.
Vellum, once a difficult commodity to find, is now widely available in
a huge array of colors and patterns. Real vellum is actually the skin
of a goat or calf, but much of what is seen in stores these days is a
version made from vegetable fibers. Vellum is the perfect weight and
translucency for making quick and pretty night-lights—it's also very
sturdy. This project features a deep turquoise blue and an olive green
color scheme for a cool, sleep-inducing color combination. Here, the
vellum is curved around and attached to a cardboard backing that
slides over any simple night-light fixture. You can trim the back plate
of your design to work with any specific fixture as needed.

PIN-PRICKED VELLUM NIGHT-LIGHT SHADES

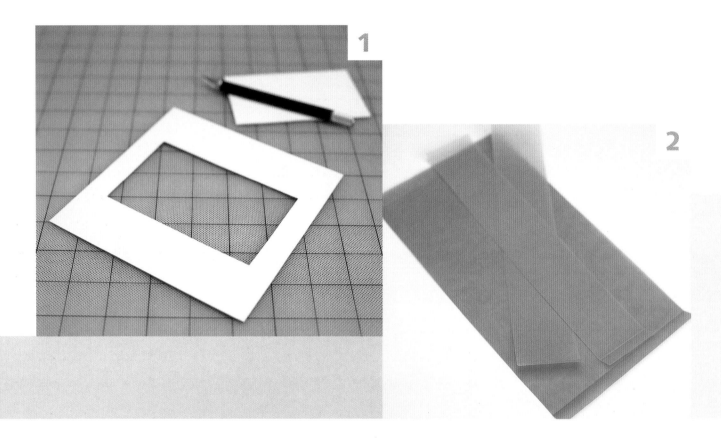

STEP 1 Measure and cut a 6" x 6" (15 cm x 15 cm) square of sturdy cardboard. Cut a switch-plate-sized hole (2 3/4" x 4 1/2" [7 cm x 11 cm]) into the center. Paint the front side with two coats of white acrylic paint.

STEP 2 Measure and cut a 6" x 11" (15 cm x 28 cm) piece from the turquoise sheet of vellum. Lay the sheet out on a flat surface. Create a 1/2" (1 cm) fold along each short side, with both folds made toward the front (top) of the paper. Cut two 2" x 11" (5 cm x 28 cm) strips of olive vellum and make a similar 1/2" (1 cm) fold on each short end of these strips, also folding toward the top.

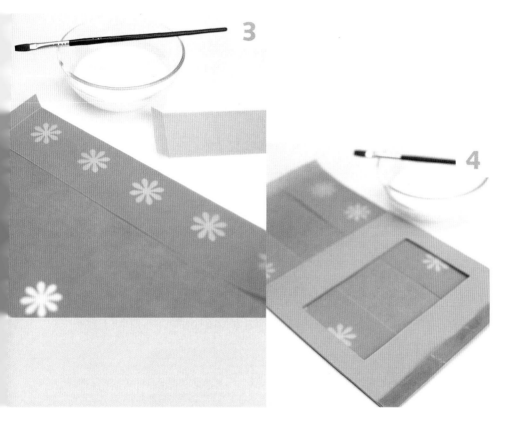

STEP 3 Punch five flowers, space evenly along the two long sides of the turquoise sheet but within the area between the two folds. Lay the sheet on your work surface with the folds facing up. Lay an olive-colored strip along the top and bottom of the turquoise sheet to back each flowered border. Nest the folds of the olive sheet with the turquoise sheet. Thin a small amount of PVA glue with 25 percent water, and then paint it into the folds to secure the sheets together. Let dry.

STEP 4 Bend the completed vellum shade into a half cylinder shape and fit the back plate into the folded edges with the painted side facing the vellum. Adhere the folds to the rear edges of the back plate using PVA glue.

VARIATION

Use a sturdy embroidery needle or any sharp tool to add pinpricked patterns to the shade. The perforations will add an interesting texture to the surface and add a dotted line of backlighting to the shade. To create a pinpricked design, lay the paper over a yielding surface, such as a small stack of newspapers. Here, the pinpricks were designed to resemble the path of a bee stopping at the flowers along his path. To re-create this design, use a craft knife to cut the bee shape from the turquoise vellum at the same time you punch the flowers. Then create the bee's looping path.

MATERIALS

- willow, grapevine, or other found straight and flexible tree/bush branches
- raffia
- waxed linen
- Japanese paper—make your own or purchase papers such as kozo, washi, or unryu
- methylcellulose or PVA glue
- socket—keyless, standard base
- 2 locknuts
- 4" (10 cm) cross bar (fixture strap)
- 1 1/2" (4 cm) to 2" (5 cm) threaded nipple (lamp pipe)
- 8' (2.4 m) cord set with plug
- cord switch
- 40-watt bulb

TIP

The paper does not require any kind of fire retardant as long as the bulb does not touch the paper. Use a 40-watt, or less, bulb and provide a heat release for the bulb. This can be accomplished by piercing a few holes in the paper near the heat source and near the top. Another option is to not cover an area within the armature.

This beautiful illuminated sculpture is made from willow branches and handmade paper. The willow is peeled and then soaked until it becomes pliable enough to work into a fluid armature that is secured with threads and raffia. Once the armature is completed, damp sheets of paper are applied to each area of the armature grid, creating a subtle patchwork of colors.

SKIN & BONES LAMP

STEP 1 Gather twelve straight willow shoots, or other similar straight and flexible branch, vine or reed. Each piece should measure 18" to 24" (46 cm x 61) in length. You can also purchase basket reed and vine.

STEP 2 Scrape off the outer bark of the harvested willow. Then soak the willow or basket reed in water for about an hour.

STEP 3 To begin the armature, bundle six willow shoots. Tie the tips together with waxed linen about 4" (10 cm) from the tip and then wrap over the waxed linen with raffia. Loosely wrap the bottom willow ends together with masking tape. Cut off any uneven ends.

STEP 4 Make five willow or grape vine rings increasing in diameter from 3" to 6" (8 cm x 15 cm). Overlap the ends of the rings, then tie and wrap with waxed lined and raffia.

STEP 5 Insert the first ring a bit below the raffia-wrapped tip and tie each willow shoot to the ring, first with waxed linen, then wrap with raffia. You may need to hold the ring in place with clamps or masking tape.

STEP 6 Next, attach a 4" (10 cm) ring for the base, and to hold the electrical works. Remove the masking tape from the base. Insert the ring about 2" to 3" (5 cm x 8 cm) from the bottom, and tie and wrap as above. Insert and secure three additional rings into the armature as pictured.

STEP 7 Install the lighting element. Begin by screwing the threaded 1 1/2" (4 cm) lamp pipe (nipple) into the lamp socket base. Next, screw the 4" (10 cm) cross bar (or fixture strap) into the pipe below the socket and tighten both ends with a lock nut. Follow the wiring instructions on the lamp socket package.

STEP 8 Place the lighting hardware on the top-side of the base ring. Secure with waxed linen and wrap with raffia. The socket should be in the center of the armature. It is important that the socket be centered so the light bulb does not come in contact with the paper skin. Following the package directions, attach the cord switch to the cord.

STEP 9 Cover the armature with paper. Make your own paper or purchase Japanese papers such as kozo, washi, or unryu papers. The paper can be dampened, then torn or cut to match the armature area to be covered. Stretch the paper taut over the willow and gently press the paper edges together. The wet paper fibers will bond to each other. Brush on a solution of methylcellulose to reinforce each seam, or use a solution of diluted PVA glue.

STEP 10 As the paper dries, it will shrink around the armature. Use a hair dryer to speed up the drying process; you will have better control and can smooth out any unwanted wrinkles in the paper.

Linda Louise Horn

Glass chimneys can be found at home centers and hardware stores in a variety of shapes and sizes. Pair one with an orphaned or antique china plate for an old-fashioned, beautiful candle holder. To create a matched set, focus on one simple element in the plate's design and draw or trace it to make a motif that can be etched on the shade.

etched glass
candle holder

starting out:

Be sure to apply a thick, uniform layer of etching creme for an evenly frosted shade. Be careful not to let any creme linger inside the shade when rinsing off the excess, or a faint etching will be visible on the outside.

1 **Create a motif for the chimney.** Determine the size of the motif to be etched by cutting out a piece of graph paper and holding it to the shade. Draw a simple motif on graph paper, using the paper's lines as a guide to create a symmetrical design. If using the china plate as inspiration, try tracing a design from the plate first. Then sketch a larger version on the graph paper. Next, cut the pattern out of the paper and trace it on a piece of adhesive vinyl using a permanent marker. Use a craft knife to cut the pattern out of the vinyl, making sure not to cut into the pattern.

2 **Etch the chimney.** Wash and dry the chimney with a lint-free rag. Firmly press the vinyl pattern to the chimney. Following the manufacturer's directions, apply a thick even layer of creme to the glass with a sponge applicator. Wait five minutes, then use the applicator to scrape off as much of the creme as possible and put it back in the bottle. Rinse off the remaining creme under warm, running water, and remove the vinyl pattern. Be sure to wear rubber gloves when working with etching creme.

SHORTCUT/VARIATION:
Use a large hour-glass-shaped hurricane shade and a bright, solid-colored plate to make a striking summertime centerpiece for an outdoor gathering. Etch a bold, easy-to-cut design on the plate and the shade, such as stripes.

MATERIALS
- glass chimney
- china dessert plate or saucer
- graph paper
- adhesive vinyl
- etchall™ glass etching creme
- sponge brush applicator
- craft knife
- rubber gloves

QUICK TIP
Since the sketched motif needs to be cut out of adhesive vinyl, keep the lines very simple.

These cutout shades sparkle when lit, and they make the most out of candlelight. Crumpling soft embossing copper gives it added strength and durability, and it also creates more surface area to reflect light. Try leaving the shades outside for a real verdigris patina. For a dappled rainbow effect, hold the copper with a glove-protected hand over an open flame.

crumpled-
copper
candle shades

starting out:

Use old scissors to cut metal, which is very dulling and can ruin a good pair of scissors. Also, change the craft knife blade often while cutting out the pattern. Save the discarded blades for other uses, such as scoring or cutting cardboard.

MATERIALS

- soft embossing copper
- fine copper wire mesh
- quick-grab adhesive
- old scissors
- hole punches
- craft knife
- fine-tipped permanent marker
- masking tape

QUICK TIPS

To make identically sized triangles, make a cardboard triangle and trace it randomly over the copper.

Metal shades may become hot if the candle has been burning for a while, so be careful when handling them.

1 **Cut out the patterns from the copper.** First, determine how large the shade should be by wrapping a rectangle of paper around the candle. Be sure there is at least 1/2" (1 cm) of clearance on all sides of the candle. There should also be 1/2" (1 cm) overlap on both short ends to close the shade. Add 1" (3 cm) to the height of the rectangle so that the top and bottom edges can be neatly folded over. Next, measure and cut the copper sheets using the paper pattern and a pair of old scissors. Then, use a sun pattern to make a stencil. Tape down the copper, and transfer the image using a fine-tipped permanent marker. Cut out along the lines using a sharp craft knife.

2 **Crumple the copper.** Gently crumple the copper, being careful not to rip the delicate areas of the sun pattern. Then, smooth the metal back out. Use a rolling pin to flatten it out.

3 **Add the wire-mesh backing.** Cut a piece of wire mesh that is 1" (3 cm) shorter and about 1/2" (1 cm) less wide than the shade. Place it over the crumpled copper, and fold the top and bottom edge of the copper over 1/2" (1 cm) to secure the mesh in place. Use the edge of a ruler for a straight, even fold. Use a rolling pin to flatten the crease. Next, punch a border along the top and bottom edges, taking into account that the short ends will overlap by 1/2" (1 cm). Star and crescent punches were used here.

4 **Close the shades.** Use a thin line of quick-grab glue to close the shades. Overlap the short ends at least 1/2" (1 cm) and press firmly along the seam for about one minute. Then, wipe away or gently rub off any excess glue.

SHORTCUT/VARIATION:

 Try making a pleated shade from wire mesh. Fold the top and bottom edges over for a clean finish, then use a ruler to make neat folds. Lace the edges shut with fine-gauge copper wire.

Materials

hanging light with glass shade

opaque and translucent stained glass

wooden craft stick

glass cutter

grease pencil

craft knife

sand paper

clean cotton rags

clear adhesive for bonding glass to glass

navy or black sanded grout

acrylic admixture

safety goggles

latex gloves

filter mask

mixing bowls for cement and grout

Glass Patio Lights

Reminiscent of stained-glass lamps, these glass patio lights can set a festive, romantic, or soothing mood, depending on the color of the tiles that cover the shade. Use only glass shades to ensure the most brilliant, sparkling light. A variety of prewired shades, such as the ones seen here, can be found at home centers and lighting stores. The easy-to-cut custom tiles used in this project are made from sheets of stained glass using one simple tool.

Artist: Linda Benswanger/Mozayiks

Since both the shade and the tiles are transparent, it is very important to use adhesive specially made for bonding glass to glass that dries clear. Ask for it at the local hardware store.

STEP 1

Wearing safety goggles and sturdy work gloves to protect hands from sharp edges, cut the glass into square and rectangular tiles using a glass cutter. First, use a ruler and a grease pencil to measure and mark the glass for cutting. Then, use firm, even pressure to score the glass with the cutter along the first line. Next, lightly tap the glass along the scoring with the end of the cutter, and gently snap the glass into two pieces. It should break along the scored line. Repeat the procedure until all the tiles are cut. After cutting the glass, it is helpful to sort the pieces by color.

STEP 2

Squeeze some glass adhesive on a piece of cardboard or a plate. With a wooden craft stick, spread a 6-inch (15-cm) area of the shade with adhesive, and spread a small amount on the back of the tile. Begin applying the tiles. Cover the entire shade, fitting the pieces together very tightly to ensure maximum luminosity and a minimum of grout. Let it dry for twenty-four hours.

Tip: Use opaque or semi-opaque glass tiles for this project, which will conceal any grout that may have settled behind them.

Tip: If the glue is runny or very wet, let it dry to tackiness on the shade before applying the tiles.

Shades of all shapes can be decorated using the same technique described for the main project. Glass tiles in pastel colors were used on this shade to create a softer look.

STEP 3

Prepare grout for outdoor use, then spread a small amount on the shade with rubber glove-protected hands. Work the grout completely into the crevices between the tiles using a circular motion. Be careful to continually wipe excess grout off the tiles. Continue until all the spaces between the tiles are filled. Once the grout has dried enough to form a powdery film on the top of the glass tiles, use a dry rag to wipe off the film. Let the shade dry for twenty-four hours, then buff the glass with a clean cotton rag.

Tip: If there are any sharp edges sticking out of the grout after the project is completed, use a medium-grade sandpaper to remove them.

This charming and elegant sconce shade made from parchment paper and a copyright-free antique ironwork design will add a vintage European look to your home décor. Applying styrene to the parchment before attaching it to a purchased wire lampshade frame adds durability to the final piece. Line a hallway with multiple shades for an architectural lighting effect, create just one for accent lighting, or, use as a nightlight for a dark corner of your home.

image-transfer
sconce shade

Makes one lampshade

1 Enlarge template from page 293 and photocopy onto parchment paper for a finished size of approximately 3 1/2" x 9 1/2" (9 cm x 24 cm). Lightly spray with sealer to prevent toner from smudging. Let dry. Cut out pattern, leaving a margin of 1" (3 cm) on all sides.

2 Cut a piece of styrene to the same size as the parchment. Peel the backing from the styrene, exposing the adhesive back. Place the parchment right side up onto the styrene and press to adhere. Hold the paper to the wire form so that the design is centered and trim shade to 3/8" (1 cm) beyond wire on all sides.

3 Brush a scant 3/8" (1 cm) line of glue on the front of one edge of the shade and apply ribbon. Repeat for remaining edges, mitering corners. Let dry for one hour.

4 Lay shade facedown and apply a generous bead of glue 3/8" (1 cm) from all sides. Set edge of frame on glue line at one short end of the shade and secure with clothespins, cushioning pins with scrap cardboard. Roll the frame along the glue lines to the other short edge, and secure as before with clothespins. Let dry overnight.

VARIATION
Search clip-art books for other ironwork designs.

MATERIALS

- 4 3/4" x 4 1/4" (12 cm x 11 cm) wire lampshade frame to fit candelabra bulb
- 8 1/2" x 14" (22 cm x 36 cm) parchment paper
- template (see page 293)
- self-adhesive styrene
- 1 yard (.9 m) 3/8" (1 cm) wide black grosgrain ribbon
- Mainly Shades quick-dry glue
- spray sealer
- spring-clip clothespins
- general craft supplies

TIP
To ensure safety, only use with bulbs of 40 watts or less.

A simple lampshade is transformed by transferring images from slide photographs to fabric using the Polaroid image transfer process. The special print film has an extended tonal range that enhances the dreamy, watercolor-like hues of the original photographs, making the final transfers glow in the lamplight. This sophisticated decorating technique is appropriate for any series of images, from landscape panoramas to architectural details. Imperfections in the transferred prints add interest to the final piece.

photo-transfer
lampshade

Makes one lampshade

1 Expose your chosen slide in the slide printer onto the appropriate Polaroid instant print film specified by printer manufacturer. The film and printer types determine the size of the print. See package information for more details.

2 Rough-cut a piece of fabric large enough to hold selected print size. Soak in warm water and place damp fabric onto a glass pane or other smooth, hard surface. Excess water should be blotted from the fabric.

3 Pull the exposed film through the rollers of the slide printer. Wait ten seconds, then peel the image apart. Quickly place the negative carefully onto the fabric. With a brayer, roll across the image 4–6 times in one direction using medium pressure.

4 Keep negative in contact with fabric for two minutes while keeping both warm by running a hair dryer evenly over the surface, testing the back of the negative with your fingers to monitor heat level. After two minutes, remove negative carefully by peeling back diagonally from one edge.

5 Allow image to dry thoroughly. Trim the fabric to the image. Using a small amount of adhesive, apply the image to the outside of a plain fabric lampshade. Repeat for remaining sides.

VARIATIONS
Embellish the transferred image with fabric paints or dyes, or stain the final fabric print with tea to age. Use larger sheets of fabric decorated with transferred images to make pillows or place mats.

MATERIALS
- developed slide film
- instant slide printer
- Polaroid Type 669, 59, 559, or 809 film
- 100% silk or cotton fabric
- 8" x 10" (20 cm x 25 cm) glass pane
- brayer
- hairdryer
- white craft glue
- plain fabric lampshade
- general craft supplies

TIPS
Peeling the Polaroid negative apart sooner than ten seconds may result in a fogged image. Using heavy pressure to roll the negative with the brayer may distort the image; too little pressure creates white spots on the transfer. Do a few test pieces to acquaint yourself with the process.

ARTIST: MARGARET TIBERIO

Used as garden luminaries, these wire and paper candleholders add a subtle glow to evenings on the patio. The hardware cloth frame is silhouetted against the paper by the candlelight, creating a luminous yet structured effect. Hardware cloth is an inexpensive galvanized steel mesh available at most hardware stores. These holders are the perfect size to hold tea candles, but experiment with other sizes and shapes to create an eclectic mix of light. You can also vary the paper—try white rice paper for a Japanese theme.

wire-mesh votives

Makes four triangle votives

1 Wearing protective gloves, cut one rectangle of 14 x 43 squares (3 1/2" x 10 3/4") from hardware cloth with wire cutters. Clip away the wire that forms one short end to leave bare prongs.

2 Bend the rectangle in thirds to create a triangular form with faces of 14 squares (3 1/2" square, or 9 cm square). Attach the short ends by looping the bare wire prongs of one end around the other, adjoining ends, and pinch them closed with pliers.

3 Cut a 3 1/2" x 11 1/2" (9 cm x 29 cm) rectangle from the paper. Mix 2 parts matte medium with 1 part water in a container. Wrap the paper around the wire form and paint thickly with matte medium mixture. Soak the paper thoroughly with the mixture, pressing it firmly to the wire form with the foam brush to adhere. Overlap the paper at short ends and press to seal. Let dry.

MATERIALS

- handmade papers, assorted colors
- 1 roll 1/4" (.5 cm) weave hardware cloth
- acrylic matte medium
- general craft supplies

TIP

Tea candles come with metal containers; use a clear glass candle base for other sized candles.

Ornamental and festive, this candleholder is made from copper foil that is thin enough to be cut with just a pair of scissors or easily embossed or pierced. The base of each candleholder is made from the circular bottom of a recycled aluminum soda can. Create multiple holders to hold groups of candles of varying heights for a dramatic centerpiece for your dinner table. For variety, make this project with other tinted metal foils such as tin.

copper candleholders

Makes one candleholder

1 Photocopy the template from page 292. Using scissors, cut a piece of copper foil 1" (3 cm) longer than the template. Tape the foil flat over a newspaper or magazine to provide a soft work surface. Tape the template over the copper and transfer the design to the metal by tracing over the lines with a ballpoint pen.

2 Cut out the perimeter of the design with scissors. Using an awl, pierce a series of holes following the interior embossed lines. Create another series of holes along the top edge of the foil if desired. Hold a ruler along the dotted fold line on the left side of the foil piece. Fold the thin edge of foil to the left of the line up and over to the center of the piece. Turn foil over and repeat for the dotted line on the right side. Interlink these folds on the short ends to form a cylinder of the foil and press to close.

3 Wearing a pair of protective gloves, pierce the aluminum soda can near the top with a pair of scissors. Cut down to the base of the can, and cut around base to separate the circular bottom.

4 Cut the series of tabs along the bottom of the project as marked on the template. Insert the upturned circular section from the soda can into the foil cylinder. Turn the tabs inside to hold the circular base in position, and apply two or three drops of super glue to secure the can section.

VARIATION

It's easy to create an aged and darkened finish on the copper: wearing rubber gloves, wipe the finished piece with a weak solution of liver of sulfur and water according to package directions, then brush with steel wool to polish.

MATERIALS

- 12" x 4" (30 cm x 10 cm) strip copper foil
- template (see page 170)
- awl
- aluminum soda can
- super glue
- general craft supplies

TIP

A large needle can be substituted for the awl.

Just Kidding Around

Nursery Lamp Shade

Even utilitarian items, *like this simple candlestick lamp, can add charm and interest to a baby's room. To make it easy, we used a self-adhesive shade that comes with its own pattern template. Self-adhesive shades are readily available at most craft stores, but you could also start with an inexpensive paper shade and simply cover it.*

The main design element in our shade is children's artwork, which we printed onto inkjet-printable fabric and fused to the shade cover. You could enlist one of baby's older siblings or perhaps a gaggle of young cousins to create the artwork for you—and to assist with the computer. And don't forget the trims—we finished the edges with giant chenille rickrack and added some shiny buttons just for fun. This colorful shade is sure to grab baby's attention whenever the lamp is switched on.

Materials:

4" x 11" x 7" (10 cm x 28 cm x 18 cm) self-adhesive lampshade

½ yard felt

1⅝ yards (1.5 meters) giant chenille rickrack

12 assorted plastic buttons

Three 8 ½" x 11" (22 cm x 28 cm) sheets of inkjet printable fabric

Three 8½" x 11" (22 cm x 28 cm) sheets of no-sew fusible adhesive

Extra-thick white craft glue

Iron and presscloth

Scissors, or rotary cutter and mat

Air-soluble marking pen

Spring-type clothespins

Instructions

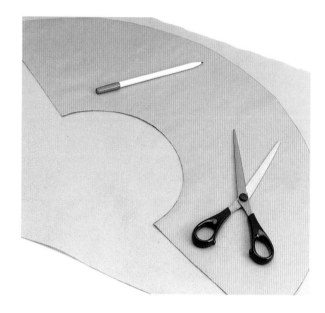

1. Remove the paper wrapper from the self-adhesive shade. Using the wrapper as a template, trace the shade shape onto felt with an air-soluble marker. Cut out the felt shade cover on the marked lines with a rotary cutter or scissors. (Note: If you are using woven fabric instead of felt, cut 1" (3 cm) beyond the marked lines all around.)

2. Scan, size, and print the selected artwork onto three printable fabric sheets. Apply fusible adhesive to the wrong side of each sheet, following the manufacturer's instructions. Cut out the images, arrange them on the felt shade cover, and fuse in place. Work from the center outward, letting some images overlap the edges as necessary.

A Word About

Design Sources

Instead of using children's artwork on a shade cover, substitute baby's photographs or some of the charming clip art images that are widely available in computer programs. You might also try a combination of the two.

3. Apply the felt shade cover to the shade, aligning the seam and top and bottom edges first. Work slowly and carefully, repositioning the cover as necessary to eliminate wrinkles. Press gently with your hand, smoothing from side to side and from top to bottom, until the cover adheres. Trim any excess fabric from the edges and seam. If the seam pops up, glue it down.

4. Starting at the seam and working a few inches at a time, apply a bead of glue around the upper edge of the shade. Glue on the rickrack so it extends slightly beyond the shade. Clamp with clothespins until dry. Trim the lower edge of the shade in the same way. Place the shade on the lamp base to avoid crushing the trim. Glue the buttons to the shade in a random arrangement.

Variation

Instead of using printable fabric art, trace some moon and star shapes onto felt. Apply fusible adhesive to the back of the felt, cut out the shapes, and fuse them to the shade cover in a random arrangement. The edging trim can be added or not, as desired.

DELICATE paper lanterns on your table and deck or along an icy walkway can add a twinkling beauty to a memorable evening. These lovely lamps are inexpensively illuminated with tea lights. By using two layers of handmade paper, one translucent and one opaque, and imaginative cutout shapes, you can make lanterns for any occasion from the Fourth of July to the winter solstice—snowflakes, anyone?

To make lanterns that rest on tabletops and decks, use 10" (25 cm) skewers; use longer skewers to make lanterns with feet that can be securely planted in sand or snow.

PHILIPPINES

SAND *and* SNOW *lantern*

Materials · 1 sheet opaque paper for outer layer, 12" x 19" (30 cm x 48 cm) · 1 rectangle thin cardboard, 6" x 10" (15 cm x 25 cm) · 1 sheet transluscent paper for inner layer, 10" x 18" (25 cm x 46 cm) · 2 strips thick paper, 1" x 18" (3 cm x 46 cm) · 6 strips of oaktag, 1" x 6" (3 cm x 15 cm) · 3 wooden barbecue skewers, 10" (25 cm) or 12" (30 cm) long · Scotch tape · pencil · ruler · craft knife with sharp #11 blade · white glue · waxed paper · 2" (5 cm) brush · scrap paper · cutting mat

TIP
For illumination, set a tea light in an empty cat food can and place inside lantern.

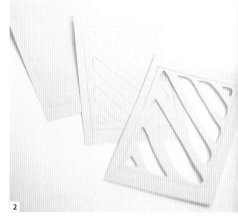

STEP 1 Fold in 1" (3 cm) top hem and 1" (3 cm) bottom hem on outer paper. Fold in 1" (3 cm) hem on one short side. Fold into three panels, 6" x 10" (15 cm x 25 cm) each panel. Place one wooden skewer in each vertical fold. Tape skewers to paper on top and bottom. **STEP 2** Cut a 6" x 10" (15 cm x 25 cm) thin cardboard rectangle. This is the finished size of one panel of the lantern. Draw a pencil line 1/3" (1 cm) in from the vertical edges and 1" (3 cm) in from the top and bottom. Draw freehand shapes (not too large) on the thin cardboard rectangle and cut out with knife on cardboard or cutting mat. Simple curves or straight lines work best. Cut out shapes to create stencil.

Choosing the Paper

Sand and Snow Lanterns look best if the outer layer of paper is darker in color than the inner layer; both are particularly effective if they have texture or inclusions (such as unryu and lace papers from Japan or botanical papers from India) that will show when illuminated. It's possible, however, to make spectacular lanterns with materials close at hand. For example, use brown grocery bags and paper doilies to produce a rustic look on a wood deck, or combine freezer wrap and colored tissue for a festive party atmosphere on winter snow.

3

4

STEP 3 Use the thin cardboard stencil to trace a pattern of shapes onto each panel of lantern outer paper. Cut out shapes. **STEP 4** To prevent paper from sticking to work surface, place outer paper on waxed paper, folded side up. Coat with white glue. Throw away waxed paper. Then place on clean scrap paper. Place inner paper on outer paper. Rub. Glue on two strips of thick paper or thin cardboard at top and bottom edges. Fold over hem of outer paper onto strips. **STEP 5** This step should be done while the papers are wet with glue so that the lantern dries into its final position. Crease folds of paper into a triangular shape. Stand lantern on edges. Glue end tab (created by the overlap of the outer paper) onto the outside surface of the opposite end of the outer paper.

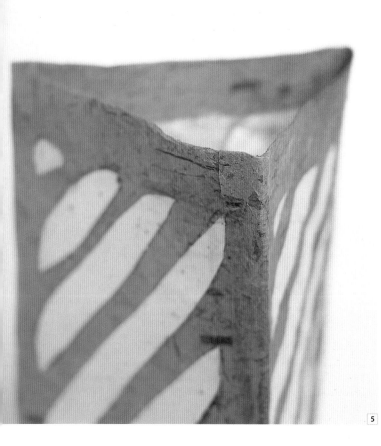

5

TIP

Consider using file folders for the lantern panels. Readily at hand and just stiff enough to stand, file folders are an ingenious substitute for thin cardboard.

About the Paper

The outer layer of the Sand and Snow Lantern is blue abaca paper from the Philippines with pieces of chopped bamboo added. Of the more than thirty different fibers indigenous to the Philippines that can be used to make paper, the most common is abaca, sometimes called Manila hemp. The whole plant is harvested and produces a long-fiber pulp that can make either thin or thick strong sheets. The West is perhaps most familiar with abaca as the material for tea bags.

The inner layer of the Sand and Snow Lantern is a type of chiri (meaning bits of bark) paper that was made in Taiwan as an imitation of Japanese paper made from kozo, a long fiber plant. In its finest form, kozo paper is thin, pure, and white; in this example, chiri remain in the pulp for decorative effect.

IMPROVISE

Achieve an altogether different shape by loosely wrapping the lantern paper around an oatmeal container. This will produce a cylindrical lantern that just fits around a cat food can and tea light. Or, instead of cutting the shapes completely out of the outer layer, consider leaving them partially attached and folding them up, down, or sideways. The candlelight will cause these "flaps" to create fascinating, dramatic shadows.

PAPYRUS is a fascinating material to use for lampshades; the overlapping strips that make up each sheet are clearly visible in the light. Further, although it appears brittle, papyrus is actually quite strong and supple, especially when wet with glue. Finally, considering the labor involved in making it, papyrus is remarkably inexpensive.

To develop an Egyptian design for your lampshade, look at books in the library. In this example, the wedja, the Egyptian symbol for "eye of Horus," was used as inspiration. This symbol means "to be whole" and is an Egyptian charm for healing.

EGYPT

EYE OF HORUS
lampshade

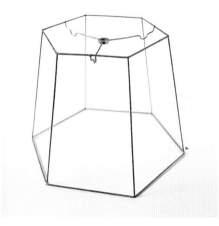

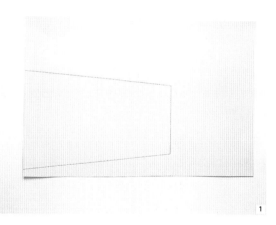

1

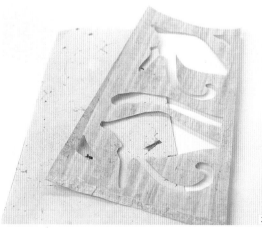

2

Materials · 1 piece oaktag or thick paper, 18" x 24" (46 cm x 61 cm) · 1 faceted lampshade frame and matching base · 3–4 pieces papyrus, 11" x 17" each (28 cm x 43 cm) · 1 sheet paper to line shade, 22" x 30" (56 cm x 76 cm) · scrap paper · white glue and brush · extra-tacky white glue with small nozzle tip · spring clothespins · scissors · knife · cutting board · pencil

GETTING STARTED To make the lampshade pattern, trace one lampshade facet (panel) onto oaktag or thick paper. This pattern should extend beyond the top and bottom of the shade by 1/2" (1 cm) and the metal frame sides by about 1" (3 cm).

STEP 1 Using the prepared pattern, cut six pieces of papyrus and six liners. **STEP 2** Using images from books on Egypt as inspiration, draw a design on the panel pattern that is interesting but also easy to cut. The design can be an allover pattern (as in the example shown here) or a frieze on the bottom and top of the shade.

TIP
It's easier to match a shade to a base than the other way around.

If the lampshade frame comes from a yard sale, clean the shade and remove any previous covering. Make sure that the wiring in the base is safe. Take it to a hardware store if you are uncertain.

Take advantage of copiers to enlarge or reduce images to suit your purpose.

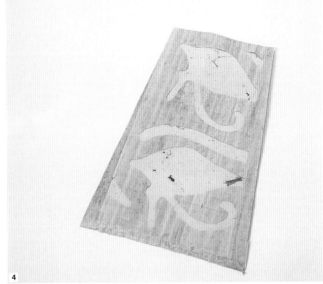

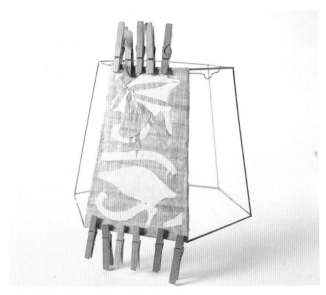

STEP 3 Cut the design out of the pattern to obtain a pattern/stencil combination. Trace it onto the six panels of papyrus. Cut the design out of each panel. **STEP 4** Brush each piece of papyrus with white glue and stick it to the liner paper. Rub with clean scrap paper to make the papyrus lie smoothly. **STEP 5** To attach the shade, spread tacky glue from a container with a small nozzle onto the lampshade frame. Lay the lined papyrus panel on top, fold top and bottom 1/2" (1 cm) over the frame, and secure with spring clothespins. Trim panel sides to edges of frame. Proceed facet by facet around the lampshade. Each papyrus facet will overlap the next.

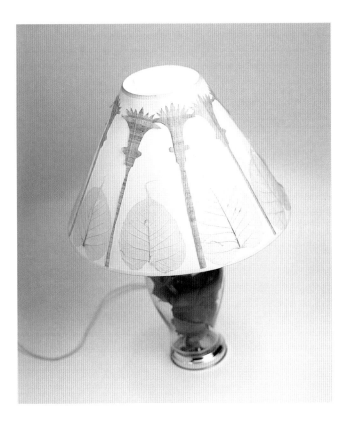

Cutout Papyrus Lampshade

Because papyrus sheets are small, it's impossible to make a large shade in one piece. As an easy option, glue cutout papyrus shapes onto a large lampshade surface. The result will be as interesting as the smaller shades.

About the Paper

Papyrus, a proto-paper, is made from the pith of the papyrus plant, which is triangular in cross section. The plant is peeled, sliced into thin strips, and soaked in water. Sheets are formed by overlapping the strips at right angles and pressing them until dry.

Papyrus was an important commodity in ancient Egypt and, after the Roman conquest in 30 B.C., the chief writing material of the Graeco-Roman world for almost 1,000 years.

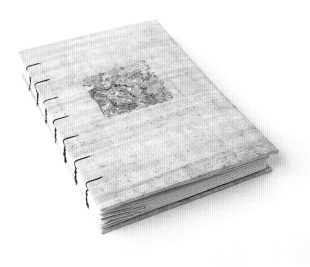

Papyrus Travel Journal

Create a travel journal using papyrus to cover the book. The papyrus becomes soft and flexible when wet, making it easy to cover the book with smooth finished edges. The unique texture and transparency of the papyrus will make this travel journal a favorite memoir or gift to someone special.

Make a quick and charming votive shade using an attractive rubber stamp and adding glitter glue for sparkle. The basic shape is made from laminated vellum and secured with curled strands of silver wire. You can use this basic technique to make shades of any size.

MATERIALS

- 8 ½" x 11" (22 cm x 28 cm) sheet of vellum
- 'zebra butterfly' rubber stamp
- black inkpad
- colored glitter glue
- hole punch
- 28-gauge wire
- votive candle

STAMPED VOTIVE SHADE

STEP 1 Stamp the butterfly in all directions all over the vellum. Let dry.

STEP 2 Apply several glitter glue colors to the wings of each butterfly. Let dry.

STEP 3 Laminate the vellum. Lay the laminated piece over newspaper and make pinpricks in a looping pattern around the surface of the shade.

STEP 4 Punch holes on the short length of each side of the laminated vellum, matching the holes up carefully.

STEP 5 Form the vellum into a lantern shape and twist lengths of wire through each pair of holes and into decorative curls to secure the edges.

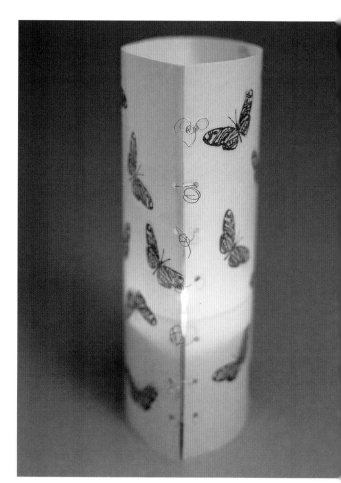

Dawn Houser

TEMPLATES

For those projects in the book that require templates, the following pages contain actual-size or reduced-sized versions. Each page containing a template will indicate what percent you need to increase it on a photocopy machine to bring it to its full size.

When making the most basic lampshade—the round empire shade—it is particularly crucial that you are working with a precise template. The template shape for an empire shade is an arc form. The mathematics to creating your own arc can be tricky. Additionally, it requires drafting tools that the average crafter does not have on hand. It is advisable to have an arc template drafted for you. Many sources for lampshade fixtures supply just this service and do so inexpensively. To order a drafted arc, you will need the diameter of the top of the shade and the base of the shade. You will also need the length of the shade's side. Of course, if you are restoring an existing lampshade, your other option is to wrap scrap or tracing paper around the shade and draw its outline to replicate the size of the arc you need.

THE BASIC SHADE

525%

BASIC BOX SHADE

352%

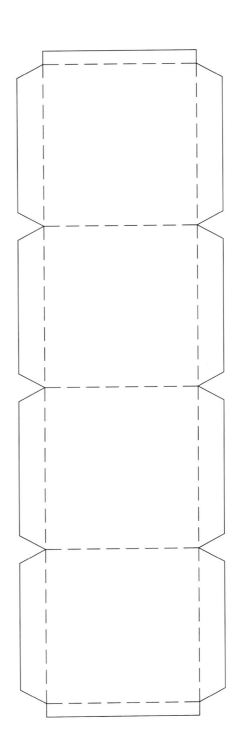

PAPER TWIST CONE

342%

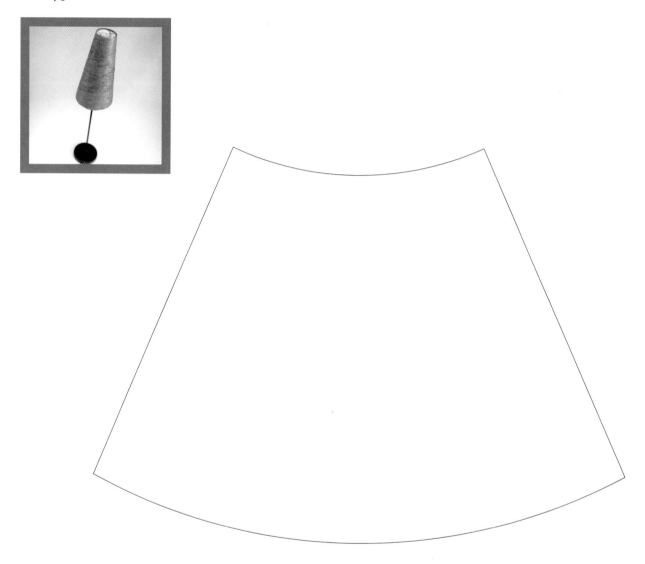

TEMPLATE 1

COFFEE-DYED LAMPSHADE
with copper lashing

248%

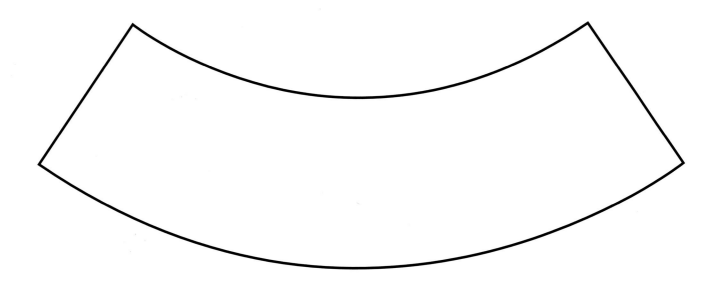

HEXAGONAL SCONCE SHADE

100% (actual size)

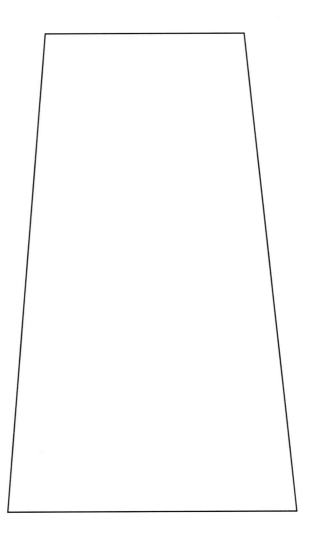

STRIPED WHIMSY SHADE

139%

CUTOUT TEA LIGHTS

148%

PINK BOA

410%

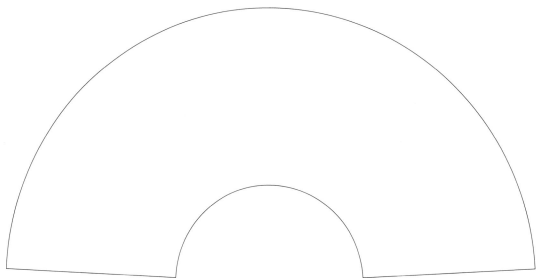

TEMPLATE 1

450%

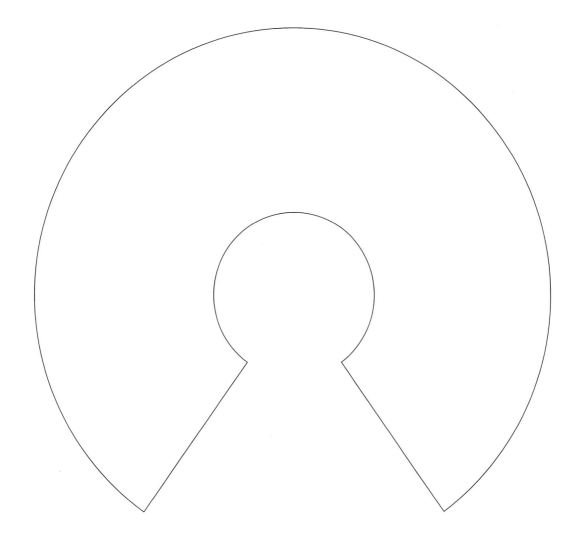

TEMPLATE 2

DRAGONFLY NIGHT-LIGHT

175%

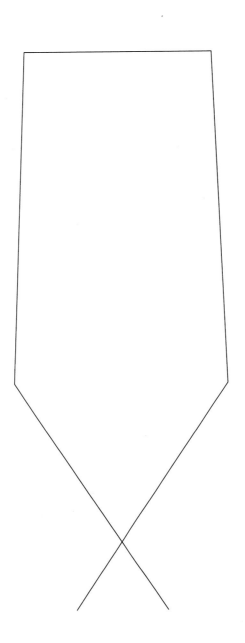

TEMPLATE 1

170%

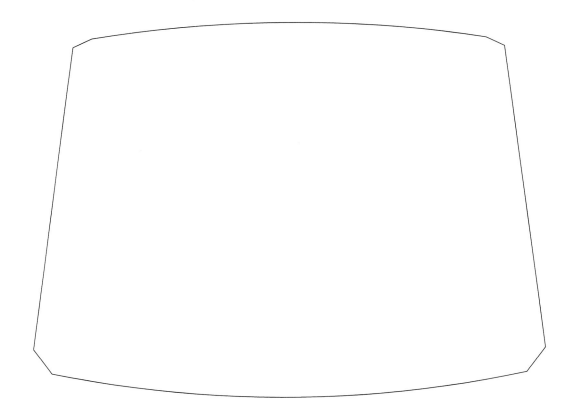

TEMPLATE 2

CRUSHED LEATHER LOUVRED SHADE

250%

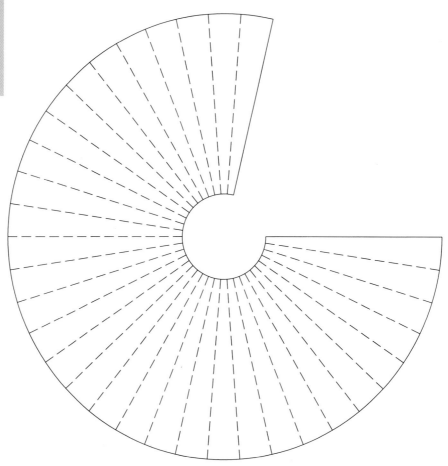